Springer Series in
Experimental Entomology

Thomas A. Miller, Editor

Myron P. Zalucki
Editor

Heliothis:
Research Methods
and Prospects

With Contributions by

P.G. Allsopp   G.S. Boyan   G.J. Daglish
J.C. Daly   V.A. Drake   M.L. Evans   G.P. Fitt
P.C. Gregg   R.V. Gunning   R.E. Jones
R.L. Kitching   D.A.H. Murray   H.E.H. Paterson
P.M. Room   J.M. Seymour   M.F.J. Taylor
R.E. Teakle   I.J. Titmarsh   P.H. Twine
A.G.L. Wilson   M.P. Zalucki

With 15 Figures

Springer-Verlag
New York Berlin Heidelberg London
Paris Tokyo Hong Kong Barcelona

Myron P. Zalucki
Department of Entomology
The University of Queensland
Brisbane, Queensland,
Australia 4072

Opposite the title page: Photograph of *Heliothis* by P.C. Gregg.

Library of Congress Cataloging-in-Publication Data
Heliothis, research methods and prospects / Myron P. Zalucki, editor.
      p.   cm. — (Springer series in experimental entomology)
   Includes bibliographical references and index.
   ISBN 0-387-97330-3 (alk. paper)
      1. *Heliothis*.   2. *Heliothis*—Research.   I. Zalucki, Myron P.
II. Series.
QL561.N7R47 1990
595.78'1—dc20
                                                                90-9865
                                                                CIP

Printed on acid-free paper.

Typeset by Bytheway Typesetting Services, Norwich, New York.
Printed and bound by BookCrafters, Inc., Chelsea, Michigan.
Printed in the United States of America.

9 8 7 6 5 4 3 2 1

ISBN 0-387-97330-3 Springer-Verlag New York Berlin Heidelberg
ISBN 3-540-97330-3 Springer-Verlag Berlin Heidelberg New York

Series Preface

Insects as a group occupy a middle ground in the biosphere between bacteria and viruses at one extreme, amphibians and mammals at the other. The size and general nature of insects present special problems to the study of entomology. For example, many commercially available instruments are geared to measure in grams, while the forces commonly encountered in studying insects are in the milligram range. Therefore, techniques developed in the study of insects or in those fields concerned with the control of insect pests are often unique.

Methods for measuring things are common to all sciences. Advances sometimes depend more on how something was done than on what was measured; indeed a given field often progresses from one technique to another as new methods are discovered, developed, and modified. Just as often, some of these techniques find their way into the classroom when the problems involved have been sufficiently ironed out to permit students to master the manipulations in a few laboratory periods.

Many specialized techniques are confined to one specific research laboratory. Although methods may be considered commonplace where they are used, in another context even the simplest procedures may save considerable time. It is the purpose of this series (1) to report new developments in methodology, (2) to reveal sources of groups who have dealt with and solved particular entomological problems, and (3) to describe experiments which may be applicable for use in biology laboratory courses.

THOMAS A. MILLER
Series Editor

Preface

As will be apparent in the title and throughout this book, I have decided to retain the more traditional generic name *Heliothis* rather than use *Helicoverpa*. Hardwick (1965) proposed the generic name *Helicoverpa* for *H. zea, H. armigera, H. punctigera*, and *H. assulta*. This revision did not receive much support for many years, although recently a PhD thesis (Matthews, 1987) has been cited as support for the taxonomic correctness of this nomenclature. As Nye (1982) pointed out, "generic boundaries are generally a matter of opinion, depending on the user's tendency to lump or to split." He suggested it was better to retain the generic name *Heliothis* for fieldworkers and use *Helicoverpa* as a subgenus for the convenience of taxonomists! (Common [1990] uses *Helicoverpa* as a subgenus in a recent work on Australian moths.) Not without precedent (see Fitt, 1989), I have used *Heliothis* to maintain consistency with the ecological and economic literature. It is unfortunate that a group of insects of such enormous economic significance has had (and apparently will continue to have) an unstable nomenclature (see e.g., Nye, 1982). This perhaps in part attests to how difficult this group of species is to study.

The origins of this book lie in a workshop on the ecology of *Heliothis* in Australia that brought together researchers from universities and state and federal government instrumentalities to review the state of knowledge concerning this problem group of species (see Zalucki and Twine, 1986). That review highlighted the lack of basic research on these species. In Australia, *H. armigera* (Hübner) and *H. punctigera* Wallengren are recorded as pests of virtually all field and horticultural crops. The pest problems have apparently increased over the years as the total area given to growing agricultural host plants has increased. Further, the mixture and sequence of crops within an area is often such that hosts are available throughout the animals' active season. Research on these pests has concentrated largely on control, often limited to a single agricultural host (such as cotton, soybean, tobacco; see Zalucki, Daglish, Firempong, and Twine, 1986). It is a telling comment on the state of research in Australian agriculture that there are no up-to-date figures on the scale and cost of the *Heliothis* problem, despite the fact that *Heliothis* are acknowledged as the major pests of field crops. Further, the continued overuse of certain insecticides has led to a high frequency of pyrethroid resistance in *H. armigera* (this is despite the widespread adoption of rational pest management practices based on the SIRATAC program in cotton). The pyrethroid problem has followed the development of DDT resistance in the early 1970s and indicates that few lessons have been learned. The resistance management program (Anon, 1983) brings into sharp focus the lack of basic understanding of the species ecology. We do not understand the spatial and temporal population dynamics of *Heliothis* spp. and are not in a position to project a priori the effect of an areawide insecticide spraying strategy. It is with uncertainty that levels of resistance are monitored and it is hoped these will not continue to rise in sprayed and unsprayed populations (Gunning, 1988). The resistance problem also serves to highlight the lack of viable, tested alternatives to management with insecticides.

It was against this background that the *Heliothis* workshop led to a call for more research into the basic biology and ecology of these species on an areawide basis (see the recommendations of Zalucki and Twine, 1986). One specific resolution stated that a handbook of research techniques should be produced by experienced *Heliothis* workers. This volume is the result.

The objective was to draw on the experience of researchers who have worked with *Heliothis* to outline research techniques and methods that others may find useful. In some instances, "recipes" that have worked in specific areas such as collecting, culturing, and resistance monitoring are presented. However, this book is not just a recipe book for research. In the main, chapter authors identify key problem areas for further work and outline approaches that will need to be tested and developed further by other researchers, depending on what question is being asked.

Research in any of the sciences does not consist of the application of the latest technique. It is essential that considerable thought be given to framing questions concerning the biology/ecology of *Heliothis* taxa. These questions are rarely formulated in a vacuum. They are often colored by the particular view of life one holds. For instance, if one believes that a population is "regulated around some equilibrium," then the question one may seek to answer is: What density-dependent factor regulates the population? Considerably more progress can be made if the view of life one holds is also brought into question. The question will then be framed so as to distinguish between alternate answers and experiments designed accordingly. Techniques or methods become important only in the latter; they should be such that they answer the questions unambiguously. Further, checks of technique must always be made to ensure that it measures what it is supposed to measure.

One of the difficulties with working on *Heliothis* is that it is not a technically convenient group of species (see Gilbert, Gutierrez, Frazer, and Jones, 1976). *Heliothis* spp. can be difficult to collect and first instars difficult to find; cultures often die out quickly; adult behavior is nocturnal and difficult to observe and movement extensive; females potentially can oviposit on a wide range of plants and so on. Many of these problems are addressed in the various chapters.

One of the difficulties with *Heliothis* spp. has always been in identifying the species with which one is working. Common (1953, 1985) described five Australian species taxa using classical taxonomic criteria — dissections of genitalia and wing characters — that are only reliable in fresh specimens. Hugh Paterson (Chapter 1) raises the question of cryptic species. The nature of the question hinges on how one defines species. The existence of cryptic species would radically change our interpretation of *Heliothis* species ecology. The techniques for testing for the existence of separate gene pools are reasonably straightforward (see Chapters 1 and 13) and should be done as a matter of high priority, before much more ecological work is done.

Collecting eggs and larvae (Peter Twine and Margaret Evans, Chapter 2) is part of many research projects. Due to the effort and cost of doing field trips to find animals, collecting should be done in such a way that mortality is minimal and maximum information is extracted. In many research projects there is a need to establish a laboratory culture (Robert Teakle, Chapter 3). Although this can be done and a procedure is outlined for this, it is not often clear why cultures fail to establish or give out after some time. Some work on this problem and on what aspects of the animals' behavior and physiology change when *Heliothis* are put into culture would be useful.

Light and pheromone traps are used by entomologists for various reasons (Peter Gregg and Angus Wilson, Chapter 4). Surprisingly little work has been done on trap design to improve performance and on interpretation of what the significance of the collection of a certain number of insects per night is.

Even less systematic work has been done on sampling techniques for *Heliothis* spp. immatures on their many and various host plants (Ian Titmarsh, Myron Zalucki, Peter Room, Margaret Evans, Peter Gregg, and David Murray, Chapter 5). The number of hosts and their varied physical and botanical characteristics are in part the problem, yet sampling the numbers accurately is essential to the construction of life tables (Peter Room, Ian Titmarsh and Myron Zalucki, Chapter 6).

Although techniques for life table construction have been around for some time, their interpretation and meaning have been the subject of much controversy (e.g., Dempster, 1983; Hassell, 1985; Dempster and Pollard, 1986; Hassell, 1987). More of the same will not resolve the issue; direct experimentation is a much better way of measuring natural enemy impact (James Seymour and Rhondda Jones, Chapter 7).

Measuring development of immatures in relation to temperature is essential in the study of any poikilotherm (Peter Allsopp, Greg Daglish, Martin Taylor, and Peter Gregg, Chapter 8). As the authors of Chapter 8 point out, this is not as simple as it sounds and many potential sources of error need to be considered. The enumeration of these sources and how (in part) to deal with them reflect the extensive work on insect development over the years. Not as much work has been done on diapause in Australian *Heliothis*. The low density of pupae in the field and the difficulty of sampling pupae in the soil have hampered studies. Several approaches using cultures and cages are outlined (David Murray and Angus Wilson, Chapter 9) and these will need to be tested further. Physiological studies on the nature of diapause are also needed.

Movement is perhaps the most difficult population process with which to come to terms (Alistair Drake, Chapter 10). The author outlines a number of different methods for detecting and measuring movement and argues for a team approach.

The study of insect behavior (Gary Fitt and George Boyan, Chapter 11) — from simply watching what the animal is doing to the neurophysiology of chemoreceptors on the female ovipositor—will greatly extend our understanding of *Heliothis* biology. There are many and various techniques here. How one perceives insect behavior (in terms of a conceptual model), as with the nature of species and population dynamics, colors how the techniques are applied and the results interpreted. The authors stress the need for field observations of behavior combined with more detailed laboratory work. The study of auditory-mediated behaviors in *Heliothis* provides a model system that could well be applied to other behaviors.

The techniques for measuring resistance may be standard (Robyn Gunning, Chapter 12), but this does not overcome the problem of providing sufficient animals to detect the true level of resistance in a population. The development of alternative methods that test at the egg or first instar stage are useful but contingent on their ability to identify the genetical species with which one is dealing.

Population genetics (Joanne Daly, Chapter 13) provides a body of techniques that can be useful in answering questions as disparate as the nature of resistance and the detection of cryptic species to the measurement of gene flow. Interpretation of results, however, is not always straightforward.

Many consider modelling to be the last step in a research program because it serves to synthesize what is known. As I have already intimated, not much is known and so perhaps modelling should be delayed. Roger Kitching

(Chapter 14) argues that there are good grounds for building models at the beginning of a project! The type, structure, and flavor of a model will depend on the question that is being investigated. Possible approaches to modelling *Heliothis* population dynamics are outlined. As with all the techniques covered in this book, the fine detail and application (use) of techniques will depend on the requirements of researchers and their projects.

This ends a brief summary of the contents and style of this book. Although throughout the authors may have slanted their thoughts toward the study of *Heliothis*, it should be apparent that much of the content is applicable to the study of many insect species.

Entomology Department   MYRON P. ZALUCKI
University of Queensland

Acknowledgments

The production of any contributed book necessarily involves the cooperation and effort of many people. Apart from the chapter authors, it is the behind-the-scenes people that make the production of such a book possible. I refer to the typists and artists who transform rough drafts to final copy, usually through many versions. I offer my special thanks to Judy Attard and Jocelyne Campbell for their special efforts in the typing of correspondence, manuscript text, and the references and to the typing staff of QDPI, Toowoomba, for retyping the entire manuscript.

I also thank Albert Hartstack for permission to redraw Figure 4.3i, the Entomological Society of America for Figures 4.3a, 4.3c, and 4.3g, Pergamon Press for Figure 9.1, Brian Hearn for Fig. 14.3, Philips Lighting Industries and Sylvania Electric Australia for Figure 4.1, and the *Australian Journal of the Entomological Society* for permission to reproduce Tables 5.1 and 5.2. All sources are fully acknowledged in the text.

Peter Twine provided much encouragement and support during the production of this book. Finally, my special thanks to my wife, Jacinta, for putting up with the out-of-hours time spent on this work. I trust it will all have been of benefit to somebody.

Contents

Contributors

P.G. ALLSOPP
Bureau of Sugar Experimental Stations, Bundaberg, Queensland, 4670, Australia

G.S. BOYAN
Zoologisches Institut, Universität Basel, Rheinsprung 9, 4051 Basel, Switzerland

G.J. DAGLISH
Department of Entomology, University of Queensland, St. Lucia, Queensland, 4067, Australia

J.C. DALY
CSIRO, Division of Entomology, Canberra, Australian Capital Territory, 2601, Australia

V.A. DRAKE
CSIRO, Division of Entomology, Canberra, Australian Capital Territory, 2601, Australia

M.L. EVANS
Department of Agriculture, Minnipa, South Australia, 5654, Australia

G.P. FITT
CSIRO, Division of Entomology, P.O. Box 59, Narrabri, New South Wales, 2390, Australia

P.C. GREGG
Department of Agronomy and Soil Science, University of New England, Armidale, New South Wales, 2351, Australia

R.V. GUNNING
NSW Agriculture and Fisheries, Agricultural Research Centre, RMB 944, Tamworth, New South Wales, 2340, Australia

R.E. JONES
Department of Zoology, James Cook University, Townsville, Queensland, 4810, Australia

R.L. KITCHING
Department of Ecosystem Management, University of New England, Armidale, New South Wales, 2351, Australia

D.A.H. MURRAY
Entomology Branch, Department of Primary Industries, Toowoomba, Queensland, 4350, Australia

H.E.H. PATERSON
Department of Entomology, The University of Queensland, Queensland, 4072, Australia

P.M. ROOM
CSIRO, Division of Entomology, Indooroopilly, Queensland, 4068, Australia

J.E. SEYMOUR
Department of Zoology, James Cook University, Townsville, Queensland, 4810, Australia

M.F.J. TAYLOR
Department of Entomology, The University of Queensland, Queensland, 4072, Australia

R.E. TEAKLE
Entomology Branch, Department of Primary Industries, Indooroopilly, Queensland, 4068, Australia

I.J. TITMARSH
Entomology Branch, Department of Primary Industries, Toowoomba, Queensland, 4350, Australia

P.H. TWINE
Entomology Branch, Department of Primary Industries, Indooroopilly, Queensland, 4068, Australia

A.G.L. WILSON
CSIRO, Division of Entomology, P.O. Box 59, Narrabri, New South Wales, 2390, Australia

M.P. ZALUCKI
Department of Entomology, The University of Queensland, Queensland, 4072, Australia

Chapter 1

The Recognition of Cryptic Species Among Economically Important Insects

H.E.H. Paterson

I. Introduction

Medical and economic entomologists are concerned with understanding the ecology of particular insect pest species, that is, their abundance and distribution. The fundamental problem that needs to be confronted is that we cannot always recognize the species we are studying. This is because of the common occurrence of cryptic species that cannot be separated by traditional taxonomic methods, no matter how carefully they are applied. For this reason, a species taxon defined purely on morphological criteria does not coincide necessarily with a species delimited by evolutionary geneticists. This impediment to scientific study can be overcome by admitting the problem and then dealing with it by bringing to bear existing genetical techniques guided by appropriate evolutionary genetical insights.

In this chapter, I briefly outline the difference between a morphological and a genetical species as well as those methods available to detect cryptic species. I also discuss some examples pertinent to (Australian) agricultural practice, with some emphasis on the species of *Heliothis*.

Central to my theme is the fact that the problem will not be solved by techniques alone, no matter how up-to-date they may be. The essential need is to be able to frame critical questions, informed by evolutionary insights. Only then do techniques become useful and important. It is often forgotten that techniques are tools that can be used by workers with skill and imagination, and what results from their use depends entirely on the conceptual grasp of the user. With this in mind, I have taken trouble to outline briefly some of the ideas needed to use available genetical tools effectively (see Chapter 13).

II. Kinds of Species

The term *species* has a number of quite different meanings, a fact which is not always well understood by biologists. As a result of this, we often do not know what kind of species is under discussion. For example, much confusion results when the word *species* is used to mean morphological species by a speaker, but it is understood by a listener to mean one defined on genetic grounds, or vice versa.

In taxonomy, we deal with the classification and naming of organisms. Traditionally, we have relied on taxonomists to identify and name our specimens. To the general benefit of biology, taxonomists have amassed our great collections, and have ordered them according to the system of hierarchical categories of Linnaeus. In at least 99% of cases, assignment of a specimen to any category from species to kingdom is done by interpreting distinct and shared structural characters. Taxonomic classification based on structure is equally effective in dealing with biparental (sexual) organisms and uniparental (parthenogenetic or asexual) ones. For example, taxonomy handles the uniparental species of morabine grasshopper *Warramaba virgo* in the same manner as it deals with its closely related biparental congeners (Key, 1976: pp. 64-66).

On the other hand, evolution is a process involving genes in populations, and is studied appropriately through the insights of genetics. In evolutionary genetics, which amounts to the population genetics of natural populations, species are necessarily recognized on genetical criteria. Fortunately, the genetical species often coincides with the morphological species, although not always. Quite commonly, a species taxon comprises two or more undetected cryptic (sibling) genetical species. These would not have been missed by the taxonomist through incompetence, but rather because the cryptic species concerned often really do look identical. If one relies on visual morphological criteria to erect a species, the conclusion is inevitable that specimens that look identical are part of a single species. The nature of cryptic species will become clear after genetical species have been discussed.

There is general agreement among evolutionary geneticists that a genetical species can be regarded as "a field for gene recombination" (Carson, 1957). This simply means that members of a genetical species can exchange and recombine genes freely with other members of the same species, but generally not with members of other species. Debate still occurs over what restricts the exchange of genes to this "field," and two explanations are competing currently.

Most evolutionary geneticists believe this is because of a class of adaptations, the isolating mechanisms, as was suggested originally by Dobzhansky in 1935 and supported subsequently by Mayr (1963). I have abandoned this "isolation concept" (Paterson, 1978, 1985) because there is little or no evidence that such characters have evolved to "protect the integrity of the genetic system of species" as supporters of the view claim (Mayr, 1963: p. 109), and because of many internal inconsistencies with the concept (Paterson, 1985).

Because biparental organisms in their normal habitats generally exchange genes only with other members of the same species, there is good evidence that they always possess mechanisms to attract and recognize mating partners of the appropriate kind as a prelude to achieving successful fertilization. In other words, a species is that most inclusive group of biparental organisms that shares a common fertilization system (Paterson, 1982, 1985).

By fertilization system, I mean all those characteristics that contribute to achieving fertilization under conditions that generally prevail in the organism's normal habitat. In motile organisms such as insects, the fertilization system always includes a signalling system involving the potential mating partners or their sex cells. I call this signalling system the *specific-mate recognition system*, or SMRS. It is the transmission of signals, their reception, and processing that enables males and females of an insect species to detect and "recognize" each other as a prelude to mating. After potential mating partners have come together, there are further exchanges of signals that are also necessary to achieve fertilization. Ultimately, egg and sperm "recognize" each other chemically. By recognize I mean recognition of the sort involved in the recognition of an antigen by an antibody; I do not wish to imply choice of any sort. The SMRS includes reproductive behavior or courtship, but is more extensive than that as it includes the exchange of signals by egg and sperm, or pollen and stigma.

This way of understanding species in terms of mate recognition is a "common sense" one, which seems to apply perfectly well to all biparental, sexual organisms from protozoa to *Homo sapiens*. It is very clear, too, that, in their normal habitats, males and females of cryptic species, which look identical to taxonomists, find and recognize each other, and mate without difficulty. Because they must be using their fertilization system to do this, it seems logical to recognize that it is this that sets the limits to the species gene pool or "field of gene recombination." Indeed, what could be more fundamental and sensible than for us to recognize species by the same system that the organisms themselves use when finding a mate?

Supporters of the isolation concept have attempted to explain cryptic species in a variety of ways (Mayr, 1942, 1963; Dobzhansky, 1951; Sokal, 1973; Ayala, Tracy, Hedgecock, and Richmond, 1974), but this has proved difficult in terms of reproductive isolation. The recognition concept of species, in contrast, elucidates the nature of cryptic species. Cryptic or sibling species, in fact, are not species that have yet to diverge more fully. They are normal genetical species in every way and cause difficulty only because human taxonomists use their optical sense in identifying species. If humans were as well provided for with other senses as insects are, species would not be cryptic. Another point that should be remembered is that it has been well established (for example, by Lambert and Paterson, 1982) that morphological resemblance is not always a reliable indication of phylogenetic closeness. Morphologically very different species are sometimes much more closely related than two cryptic species may be (for example, *Drosophila silvestris* and *D. heteroneura*: Johnson, Carson, Kaneshiro, and Steiner, 1975). Because cryptic or sibling species use mostly nonvisual signals in their specific-mate recognition systems, we can expect cryptic species to occur most frequently in groups that live under conditions where visual signalling is inefficient, or where visual signals are transient and cannot be preserved in a museum specimen. This point is well made by comparing diurnal Lepidoptera with nocturnal ones. Butterflies are diurnal and rely mainly on visual signals in finding mates (Tinbergen, Meeuse, Boerema, and Varossieau, 1942). Similarly, there are diurnal moths that utilize optical signals in their SMRSs, and these are colorful. By and large, there is a good level of coincidence between taxonomic species and genetical species in such groups. The difficulties arise in such groups as the Noctuidae and Geometridae, which are nocturnal and, therefore, rely largely on chemical and tactile signals. It is in such groups that cryptic species are common. In fireflies, on the other hand visual signals occur though these insects are nocturnal. Of course, the signals are transient and of little use in general museum taxonomy. In contrast, to the evolutionary geneticist they are critically important, as they are to the organisms themselves. To sum up: cryptic species are likely to be found in any group of organisms that use mainly auditory, chemical, tactile, or transient optical signals in the critical, discriminating stages of their SMRS.

III. Practical Methods for Detecting Cryptic Species

The existence of cryptic species is, ipso facto, often unsuspected. First, I discuss ways by which they have been disclosed most often in the past.

Most commonly, one comes to suspect the existence of a species complex from the occurrence of biological discontinuities, either local or geographical. For example, the genetical species, *Anopheles melas* and *An. merus*, two cryptic species, were detected within the species taxon *Anopheles gambiae* through their ability to breed in highly saline waters in West and East Africa, respectively. This is quite unlike other populations of the species taxon, which breed only in freshwater (Muirhead-Thomson, 1948, 1951, Paterson, 1962), or, at most, in mineral waters. The species taxon, *Perthida glyphopa* occurs as a leaf-mining parasite of Jarrah (*Eucalyptus marginata*), and Common's type material was limited carefully to specimens from this host. However, morphologically very similar specimens are found on Flooded Gum (*E. rudis*), and on Prickly Bark (*E. todtiana*). Using electrophoretic and distributional studies, Mahon, Miethke, and Mahon (1982) provided evidence for the view that distinct genetic species occur on Jarrah and Flooded Gum, and that the specimens on Prickly Bark may form also a distinct, third, gene pool.

With the increasing interest of applied biologists in the use of sex pheromones in the biological control of insect pests, it is not surprising that evidence for cryptic genetical species has emerged fortuitously (Carde et al., 1978), though such evidence is not always interpreted by the finders correctly.

Interesting work by Whittle, Bellas, Horak, and Pinese (1987) can be cited to illustrate how sibling species may be detected through biological clues. A tortricid moth species, *Homona spargotis*, a pest of Avocados in northern Queensland was synonymized with a Sri Lankan coffee pest, *H. coffearia* on morphological grounds. However, pheromone analysis revealed significant differences in this important constituent of specific-mate recognition systems between the Sri Lankan and Australian moths. This led to the discovery of small but consistant morphological differences.

While studying the inheritance of dieldrin resistance in a northern Nigerian population of the species taxon, *Anopheles gambiae* s. str., Davidson (1956, 1958) crossed it with a susceptible strain from Lagos. Incidentally, he found that the F_1 males were sterile, but attributed this to a pleiotropic effect of the resistance allele. In the latter paper, he rejected mistakenly the possibility that two species were involved, as was suggested by Holstein (1957), in favor of his original conclusion. Much later it became evident that the two strains were from distinct genetical species, respectively, *An. arabiensis* and *An. gambiae*, s. str., and the sterility observed was hybrid dysgenesis. The correct interpretation of evidence depends on a detailed understanding of the genetics of species. In widespread species, differences in such nonreproductive characters as the ability to enter diapause may provide pointers to the existence of cryptic species also.

Evidence of broad polyphagy in a species taxon might indicate that a complex of cryptic genetical species is involved. In other words, each cryptic species might account for part of the wide range of hosts exploited by the species taxon concerned. For this reason, the possible existence of cryptic species should be investigated in any critical study of the host range of an apparently polyphagous species, particularly when the species taxon is cosmopolitan. It is by no means clear, for example that populations of *H. armigera* around the world represent only one genetical species. Preliminary studies (Daly and Gregg, 1985) have not settled this problem altogether.

It is naive to cite changes of host relationships as evidence for the adaptability of an insect species to new hosts, unless it has first been demonstrated that only one genetical species is involved. There is much misleading information in the literature that suggests that many insect species are very variable in their habits. Such evidence should be treated with caution in the absence of a detailed and competent examination of the populations to demonstrate that they are indeed conspecific and not complexes.

Indeed, host relationships can provide evidence for the existence of unsuspected cryptic species among the host plants themselves. For example, programs aimed at controlling *Lantana camara* have been complicated by the discovery that the species taxon comprises a complex of several forms, some diploid and some tetraploid (Smith and Smith, 1982). The stimulus that led to the revelation of this complex was the study of lantana as a pest and a target for biological control.

During the past 20 years, genetical and cytogenetical tools have come into use but their utility is much dependent on evolutionary theory. They can be very helpful when used by a specialist evolutionist, but they can prove to be useless or even a handicap in the hands of the inexperienced. The interpretation of results from an electrophoretic study or a cytogenetic study is not easy or obvious. Useful answers are obtained only when critical questions are posed at the planning stage. Generally, these tools will not reveal species automatically; workers planning to use them should understand that the technical ability to run gel electrophoresis or to undertake chromosome studies is only the beginning: the difficult part is the interpretation. The questions to which one seeks answers arise from the genetic theory of species: different genetical concepts dictate different questions, although some may be common. For example, both the isolation concept of Mayr and Dobzhansky, and the recognition concept accept that a species is a field for gene recombination. Thus we might ask if the species taxon *H. armigera* at Narrabri constitutes one or more fields of gene recombination. A start might be made by sampling both sunflower and cotton taking care to avoid biases. If each individual sampled is scored then for, say, 15 enzyme systems using polyacrylamide gel electrophoresis, we will be in a position to answer the following question: Are these data consistent with the view that they were drawn from a single randomly mating population? If they are statistically homogeneous, then the answer would be yes, and it could be concluded that no reason had been found to doubt that only a single field for gene recombination had been sampled. However, a decision could not be made if no polymorphic characters had been included. If all the individuals from cotton were homozygous for an allele at a particular enzyme locus, and those from sunflower were all homozygous for an alternative allele, one would be forced to conclude that two distinct fields of gene recombination were involved. It might be thought that if the frequencies of alleles at a particular locus are in Hardy-Weinberg equilibrium, then we will have evidence that the sample examined was drawn from a single panmictic (randomly mating) population. However, this is not necessarily so. Suppose two undetected cryptic species, in fact, are present, and that both the alleles at this locus are in Hardy-Weinberg equilibrium. A random sample from this mixture of two populations will be found to be in Hardy-Weinberg equilibrium also. Thus care is needed when using such evidence for inferring that one is dealing with only a single genetical species.

Chromosome cytology can be useful in the detection of cryptic species. In many families of Diptera of economic or medical importance, polytene chromosome studies can be of great value. For example, such chromosomes from the larval salivary glands are used in detecting cryptic species in Simuliidae and *Anopheles*. Polytene chromosomes from the ovarian nurse cells of adult females in some species of the genus *Anopheles*, and those from trichogen bristles are valuable in the genera *Calliphora* and *Lucilia*. In other orders of Insecta, one must rely on mitotic and meiotic chromosomes that provide much less detail, even when banding techniques are used. Nevertheless, the chromosomes can still be used to track gene flow as was done by Moran and Shaw (1977) in studying a parapatric zone in the grasshopper genus *Caledia* in Queensland. Karyotypes of related species are often indistinguishable which means that the chromosomes are sometimes useless as markers to study gene flow. Sometimes it is possible to use morphological markers to study gene flow between populations. However, care is needed to study gene exchange using structural markers, and laboratory crosses and backcrosses should precede their use in the field (Paterson, 1956). Without such preliminary studies, the finding of supposed hybrids in the field generally should be treated sceptically. Frequently, aberrant individuals are considered confidently to be hybrids on the basis of no evidence. Morphological markers are not very useful in studying cryptic species because, generally and by definition they look very much alike.

IV. Discussion

Bearing these points in mind, it can be understood readily that hitherto underappreciated problems face entomologists, who are concerned with the biological control of weeds and the pests of major crops, as well as insect ecologists specializing in host/predator, and host/parasite relationships. Sibling species among such important parasitoid genera as *Trichogramma* are scarcely touched on today. In biological control programs, great care is taken to test prospective controlling species for host range, host specificity, and so on. Yet, how much care is taken to ensure that a complex of species is not treated as a single species, or that one sibling species is studied while it is believed to be another? Very few geneticists or entomologists are trained adequately to detect cryptic species. This is a specialized field with few specialists in it.

To illustrate the problems that can result from not detecting cryptic species in an exotic parasitoid on which a control program is to be based, the case of the braconid, *Chelonus texanus* can be cited. This was imported into South Africa and released in large numbers in an attempt to control the Karoo Caterpillar, *Loxostege frustalis*. The following quote is from the account by Annecke and Moran (1977, p. 139):

> The first release of *C. texanus* commenced in October 1942 and the last consignment was despatched to the Karoo from Pretoria, according to Bedford (1956), on 1 May 1952. During this decade of production of *C. texanus*, more than 8.5 million parasitoids were released . . . against five host species in various parts of the country. In the final year of production, 1 698 000 parasitoids were transferred in the mass-rearing room directly into shipping containers by means of an ingenious light trap designed by Bedford. . . . The parasitoids were sent for release to a total of 491 sites in the Karoo and Orange Free State, and in the two or three years preceding 1950, collections of Karoo caterpillars were made . . . in areas where *C. texanus* had been released. These collections yielded about 2% parasitism by the indigenous *C. curvimaculatus*, a species not then known to attack the Karoo caterpillar, and this species was mistaken for *C. texanus*, which it resembles. According to Bedford (1951), the first of these *Chelonus* was collected in 1943 and reported on (Ullyett, 1944) as *C. texanus*; this error in identification was not corrected until February 1951. The recoveries of the misidentified *C. texanus* were a source of encouragement for the entomologists concerned: Tardrew . . . wrote "This (2% parasitism) may not seem very good, but it is encouraging, showing that the *Chelonus* (*texanus*) can survive the dry conditions of the Karoo, and that it is becoming adapted to its host" . . . Having satisfied himself of the long-standing misidentification of the indigenous *C. curvimaculatus*, Bedford recommended in March 1951 that the mass production programme of *C. texanus* be discontinued at the end of April. His recommendation was overruled, probably for reasons of political expediency, and he was requested to proceed with production and liberation for a further year. . . . In fact none of the parasitoids ever became established and finally, in May 1952, the last liberations of *C. texanus* were made, and the programme was terminated.

It may be thought that this is merely a dreadful example of the way things were once done, but I am not aware of any similar program today that routinely relies on the criteria of genetical species for identification. I believe all still depend on the findings of morphological taxonomy, despite its proven limitations. The apparent discovery of the Oriental Fruitfly (*Dacus dorsalis*) in Australia is a recent example of the sort of problems cryptic species can cause (Drew and Hardy, 1981).

Finally, I should like to conclude with the observation that I can think of few advances in insect population biology, basic or applied, that will lead to a greater increase in efficiency than a more careful attendance to the problems arising from cryptic species. By greater efficiency, I do not mean merely the saving of large sums of money across the world, but I mean also that the time of highly trained scientists will be wasted less frequently, the literature will be cluttered up less often with uninterpretable or gravely misleading results, and programs against pests will succeed more often. Detecting sibling species is not just an academic luxury; it is often the difference between success and failure, effectiveness and bumbling.

V. Conclusion

It may be appropriate to make some comments on gaps in our knowledge of *Heliothis* in particular. In order to make critical studies on these organisms it is necessary that we be certain we are studying a single genetic species and that we know exactly which species this is. In the case of *H. armigera* I do not believe we can say that we have done much to ensure that we have met these strictures.

First, a neotype should be designated if the type of Hubner's *Noctua armigera* is, indeed, lost. Evidently, the type was from Europe. Designating a neotype will specify a particular population for reference purposes. Care should be taken then to characterize members of this population structurally, electrophoretically, and biologically. Once this is done, and not before, it will be possible to recognize the typical form wherever it occurs across the world. Similar studies need to be done on the populations at the type localities of *H. armigera conferta* (Auckland, New Zealand) and, less urgently, on that of *Helicoverpa armigera commoni* (Canton Island, South Pacific), which are treated as subspecies of *H. armigera* by Marsh (1978). This is necessary groundwork to provide a firm base for the naming of populations across the world. As pointed out above, taxonomic decisions are not the crucial ones, and neither are electrophoretic ones except in special circumstances. Ultimately, it is the fertilization system under natural conditions that is crucial. With allopatric populations, the best that can be done is to test for assortative mating under conditions that make available simultaneously both forms under test in large cages.

These stringent requirements may be treated with impatience by some. However, if they are not met insecurity will persist. In Australia at present, we do not know really whether what we call *H. armigera* constitutes a single genetic species or a complex. Furthermore, we cannot be quite sure that the true *H. armigera* of Europe is present in Australia. It is also not certain that literature purporting to refer to *H. armigera* in other countries actually refers to any population in Australia. To some this may sound an extreme statement not to be taken seriously. However, I believe that in critical terms it is an accurate reflection of the true situation. At present, we rely on taxonomic decisions that we need to treat with caution, as has been demonstrated with the analysis of such other taxa as *Anopheles gambiae*.

The situation with *H. punctigera* is somewhat less difficult, because it is an endemic species. However, it, too, should be studied in detail at its type locality (Sydney) to provide a sound basis for detailed studies across the continent.

Acknowledgments. I am pleased to acknowledge assistance I have received from Drs. Rachel McFadyen, Bill Palmer, Gimme Walter, David Yates and Myron Zalucki, and, particularly, Ian Common.

Chapter 2

Collecting Immature and Adult *Heliothis* for Distribution, Host, and Parasite Records

P.H. Twine and M.L. Evans

I. Introduction

During the last 30 years, many workers have collected parasite distribution information and made host plant records for *Heliothis* (Stinner, Rabb, and Bradley, 1977; Stinner, Bradley, Roach, Hartstack, and Lincoln, 1979). Despite this work, there are still gaps in our knowledge, and such work will continue for many years to come.

Techniques used to study these aspects of *Heliothis* ecology are varied, many being specifically developed for the task in hand. Choice of techniques usually depends on the objectives of the study, the time and resources available for sampling, and the precision required if, for example, population estimates are required (see Chapter 5) or if presence/absence is sufficient. These points must be considered carefully prior to embarking on the study.

This chapter aims to give guidelines for anyone starting out to obtain data on parasite distribution and host plant associations for *Heliothis*. Sections deal with when, where, and how much to sample; sampling procedures (advantages and disadvantages of techniques are highlighted); what information to record for specimens; how to handle specimens after they are removed from the plant and before they arrive in the laboratory; rearing and storage of specimens in the laboratory.

II. Starting Out

The general geographic area to be sampled will be determined by the aims of the study, considered in conjunction with known distribution maps for the species (e.g., Zalucki, Daglish, Firempong, and Twine, 1986), and an understanding of the vegetation types (including crops) in the areas of interest. Within those boundaries, it is a question of exactly when and where it will be most profitable to sample.

For adults, the solution is relatively simple. Decide on the maximum number of traps practicable, place traps over the area (e.g., random grid pattern, fixed numbers in designated habitats, or wherever someone is willing to look after them), and clear traps regularly and as often as possible over the period of the study. Any supplementary field sampling can be determined by peaks in trap catches.

To reduce time spent clearing traps, changing pheromone caps and so forth, it is good to set up a network of local co-operators. Police, Department of Agriculture personnel, stock inspectors, and station owners or their spouses are suggestions of some people to approach (schools teachers might also be a possibility, but probably would not be able to clear traps over school holidays). To keep co-operators informed and interested, they should be visited when the traps are set up in order to have the project and procedures explained to them (and be supplied with all necessary containers, labels, postage, etc.). After this, they should be visited at least once or twice a year and given updates (this can be incorporated with trap maintenance, survey trips, etc.). Regular phone contact should be maintained, and newsletters may be appropriate also, particularly with long-term studies.

For eggs, larvae, and pupae the situation is not so easy, as intensive sampling is only profitable when the stage(s) of interest is/are known to be present in reasonable numbers. The seasonal distribution of *Heliothis* species differs from district to district and year to year. In general, *H. punctigera* are common in the spring months, whereas *H. armigera* occurs on a number of cultivated hosts during summer and autumn. *H. assulta* and *H. rubrescens* occur at various times from spring through to autumn, and no data are available for *H. prepodes*. Distinct generations do occur in most areas and can be a useful guide for timing of sampling.

On a finer scale, host records can help in decision making. Host records are available (Zalucki et al., 1986) which indicate that *H. armigera* predominates on graminaceous type hosts, *H. punctigera* predominates in the number of legume and Asteracae species acting as hosts, *H. assulta* is restricted to the *Physalis* genus, the host range of *H. rubrescens* is not well known, and larvae of *H. prepodes* have yet to be found (an adult of each of these species is shown in Figure 2.1).

Heliothis females commonly lay eggs over, in, and around growing points, reproductive parts, or on expanding leaves. Larvae will feed on vegetative tissue, but successful development to the pupal stage seems to depend on reproductive tissue forming a reasonable proportion of the larval diet. For these reasons searching on individual plants is likely to be most profitable in the vicinity of reproductive parts and growing points.

Despite knowing all this information it is still difficult to choose specific

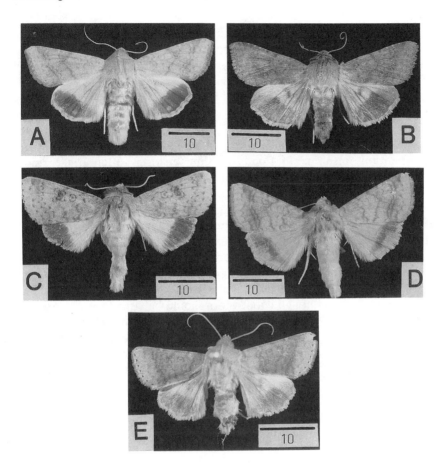

Figure 2.1. Adult specimen of Australian species of *Heliothis*: A, *H. armigera*; B, *H. punctigera*; C, *H. rubrescens*; D, *H. assulta*; E, *H. prepodes*. Scale lines are all 10 mm. (Reproduced with permission of the *Australian Journal of Zoology* from Zalucki et al., 1986.)

sampling sites, and it is probably advisable to have some set sampling sites and save some resources for ad hoc sampling (e.g., paddock that a fellow researcher has found to be full of larvae, survey on sides of roads, or every third field of cotton). As experience is gained, more profitable permanent sites may be found and some of the old sites dropped.

Sampling can be done on a set basis, but this is time-consuming and not really warranted except for population dynamics studies. If this approach is followed, sampling should be done at least 1–2 times a week to ensure all stages (particularly eggs and young larvae) are observed (see Chapter 6). The time between sampling dates may be modified depending on findings on each sampling date. Some form of sequential sampling might be employed also (e.g., if in the first 10 min, fewer than 10 eggs are found, discontinue sampling) to prevent wasting time when only low populations are present.

Plant growth stage can give some guide to sampling, as *Heliothis* adults prefer to oviposit on flowering plants. Early sown crops may be the greenest plants in an area and so attract ovipositing moths despite the fact that the plants are not flowering. Satellite imaging can also indicate occurrence of new vegetative growth that might be attractive to ovipositing moths. Weather conditions resulting in floods or thunderstorms often trigger moth flights, and oviposition on subsequent plant growth.

Another starting point is to monitor adult flights. This can be done by setting up traps at strategic points, or by relying on local citizens at those sites to let you know about large numbers of moths coming into lights. Light traps are probably most troublesome as they require a power source and more maintenance than do pheromone traps. On the other hand, pheromone traps will not give an indication of female egglaying status (see Chapter 4). The network of locals (as described earlier) can be employed also here for clearing traps, although cooperators should be visited (or at least contacted) each time you are sampling in the area. If relying on local observation of moths coming to lights, it will probably be necessary to telephone regularly (perhaps once a week) to ensure adequate reporting. (Moths from flights should also be collected and sent to you as quickly as possible for identification and assessment of egglaying status of females.)

In general, you should aim to collect at least 50 individuals for host record samples. In practice, the number of individuals will depend on population densities and what can be collected within the time available for sampling.

III. Sampling Techniques

1. Adults

The most commonly used technique for collecting adults is the light trap (see Chapter 4). Adults can be kept either alive (collecting container needs to be large, with quantities of paper or other absorbent material in it to minimize damage to moths) or dead in the collecting container (Dichlorvos-impregnated strips remain active for up to 4 weeks). A major problem with light traps is the simultaneous collection of other such insects as beetles, which can damage collected moths badly even where a killing agent is present in the collecting container. Various modifications can be made to exclude such species from traps (Common, 1959; White, 1964), and frequent trap clearances can help to alleviate this problem also.

In situations where only male moths are required, pheromone traps can be used if the appropriate pheromone is available (at present only for *H. armi-*

gera and *H. punctigera*). As for light traps, adults can be kept alive or dead in the collecting container. Details of pheromone trapping can be found in Chapter 4 and in Hartstack, Witz, and Buck (1979a,b). Pheromone trapping ensures good quality moths, but adults collected in this way should be examined to ensure they are the species expected, as occasionally an adult of another species can be collected. With this method there is no way of obtaining information on the egg-laying status of females.

Trapping allows large numbers of adults to be collected with minimum effort and clearing does not require great expertise. Moths can be drawn into traps from large distances away, and thus trapping is difficult to use for assessing populations or collecting individuals from specific habitats or small areas.

The success of both light and pheromone trapping is dependent on a number of such factors (see Chapter 4) as trap design, placement, height from the ground, and weather conditions (especially moon phase, temperature, wind speed, direction, and humidity). These factors and the necessity of keeping accurate records of them should be considered before beginning the study. Additionally, the presence of females producing pheromones may have an effect on trap catches, but this factor is difficult to assess.

Alternatives to trapping techniques are the use of sweep nets and butterfly nets. Small numbers of adults can be collected during the day by sweeping vegetation, but this is usually only worthwhile where populations are very high. Alternatively, at dusk and into the night (using car headlights or a head-mounted lamp as necessary) individual moths can be pursued and netted. This technique can be used also during the day, relying on operator movement to disturb moths, but is not usually as successful as the evening/night collection time. To minimize damage to the operator, note should be taken of obstacles and holes in the collecting area prior to commencement. A bait such as a honey solution can be placed on plants, but this usually results in unacceptably sticky nets, moths, and operators. Although numbers thus collected are usually lower than those from trapping and require more effort per unit catch, this technique can be useful when moths from particular habitats or small areas are required.

2. Eggs

Eggs can be collected by direct observation of plants only. They are found usually on the terminals, young leaves, buds, flowers, and fruiting bodies in the upper canopy of the plant. *Heliothis* eggs can be confused easily with the eggs of a number of other such noctuid species as false loopers, but with experience it is reasonably possible to differentiate these reliably. *Heliothis* eggs are laid singly, are an opaque creamy white-to-yellow (sometimes tinted green), turn grey to black prior to hatching, and do not have blotchy patches of color. When looked at from a side angle, the eggs are almost spherical, and the base is narrower than the widest diameter of the egg (Fig. 2.2A). In the center of the upper surface of the egg there is a raised boss from which striations radiate outward (some magnification may be necessary to see these characters).

Plants can be examined either in the field or cut and taken back to the laboratory. The latter option will not always be practical (e.g., for very large plants), but does allow larger areas to be covered in the field in a shorter time. Examination in the field is often easiest as plants are turgid and easier to handle, and eggs are less likely to be knocked off the plant, damaged, or hatch prior to laboratory examination.

When eggs are reared individually, it is usual to remove them singly or in small groups with a small amount of plant material still attached. This can be done by tearing the plants by hand, or by using a disc cutter (Hoffman, Ertle, Brown, and Lawson, 1970). This instrument is easy to use and can be made from scrap pieces of hardware, but does not work well once plant material has lost its turgidity. In the case of corn silks, the silks with attached eggs can be collected and returned to the laboratory. Eggs can be dislodged by moving the silks around in a container, but damage to the eggs can result and it may be preferable to cut off small sections of the silk with eggs attached.

Where data from individual eggs are not required, plants (or quantities of plant material) with large numbers of eggs present can be collected and incubated in appropriate containers. This is particularly appropriate for corn silks and sunflower heads, where large numbers of eggs are laid often. One problem with this method is the possibility of eggs of unwanted species being present.

3. Larvae

General characteristics of larvae to look for in the field are

(1) four pairs of prolegs;
(2) spines on raised bases, and
(3) the surface of the skin covered in very fine spinules (Fig. 2.2B,C,D).

Techniques used for collecting larvae include direct observation and hand picking, ground cloth (or beating), and sweep netting (see Chapter 5). Direct observation is used most widely, as it allows most information to be collected for each larva (e.g., plant part, height in the canopy), damages larvae least, and collects larvae, which are attached to the plant due to parasite emergence and pupation. However, this technique is time-consuming and sampling large areas is difficult.

Sweep netting allows large areas to be sampled quickly, and could be used to determine whether or not direct observation sampling is worthwhile, and in which spot it might be most profitable. The ground cloth can be used if

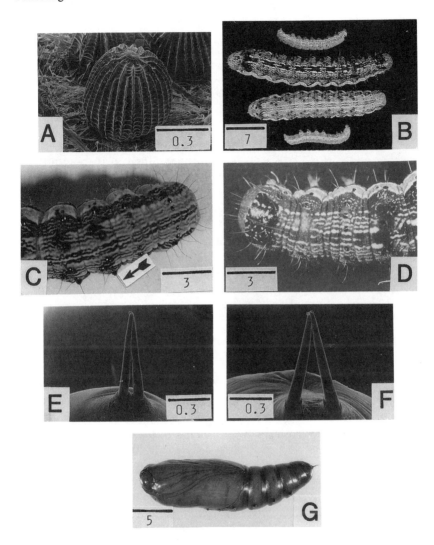

Figure 2.2. Immature stages of Australian pest species of *Heliothis*: A, egg of *H. punctigera*; B, third and final instars of *H. armigera* (above) and *H. punctigera* (below); C,D, head and thorax of *H. punctigera* (C) and *H. armigera* (D, note that setae are white); E,F, cremaster spines of *H. punctigera* (E) and *H. armigera* (F); G, pupa of *H. punctigera*. Scale lines: A, E, F, 0.3 mm; B, 7 mm; C,D, 3 mm; G, 5 mm. (Reproduced with permission of the *Australian Journal of Zoology* from Zalucki et al., 1986.)

large numbers of larvae are required and information about individuals is not important. Netting would not usually be suitable for this purpose, as it does not allow for the collection of larvae from the whole canopy, tends to collect large larvae preferentially, and causes unacceptable levels of damage to larvae.

4. Pupae

Collecting pupae (Fig. 2.2E,F,G) (see Chapters 5 and 9) is more difficult than collecting *Heliothis* at other stages. This is due to low natural pupal densities, the large quantities of soil that therefore need to be searched, and the care needed to ensure minimum mechanical damage to the pupae.

Searching in cultivated crops known to be infested with reasonable densities of final instar larvae is most profitable. In general, pupae are found within 10 cm of the planting row and within 10 cm of the soil surface. The depth of pupal cells is determined by such factors as soil type and soil moisture content at the time of pupation (see Chapters 5 and 9).

IV. Records

Adequate, uniform, and accurate records must be kept in an orderly manner if a study is to be successful. Record keeping begins in the field, and recording forms should be made up and photocopied ready for use. These forms should contain either suitable spaces for laboratory-rearing information, or separate forms should be devised. Using such forms ensures that all information is collected, and that each sheet can be filed for ease of later access. As it is likely that large quantities of data will have to be handled, a database program is the most appropriate method for permanent storage of information, and record forms should be devised with this in mind.

Information to be collected will vary from study to study, but the basic field information that must be obtained includes the date, collector's name, location of the collection site (latitude and longitude), soil type, and habitat type (e.g., roadside, sunflower crop). Good quality maps used together with the car odometer can locate collection sites to within about half a kilometer, but if exact locations (to within meters) are important, then handheld satellite navigational instruments for this purpose are now available.

Additional field information that should be collected includes insect growth stage, plant growth stage, an estimate of *Heliothis* density in the area, and an indication of the extent of larval feeding. The details to be collected will depend on the aims of the study, for example, if parasitism is of interest, then insect growth stage records should be detailed (for example, whether eggs are at the white, brown, or black ring stage), and if host plant associations are of interest, then botanical specimens or photocopies of plants/plant parts may be required.

Identification of the host plant and the *Heliothis* species being dealt with are important. If you do not have botanical expertise, then a botanist or other experienced persons should be consulted to determine the collection and preservation procedures necessary if botanical specimens are to be identified accurately. There are a number of taxonomic features (see Chapters 1 and 13) that partially define the species involved. Where distinctions are to be made between *H. armigera* and *H. punctigera*, there are keys available for eggs, larvae (Cahill, Easton, Forrester, and Goodyer, 1984; Zalucki et al., 1986) and pupae (Kirkpatrick, 1961) (see also Figs. 2.1 and 2.2), although these keys can be difficult to use. In parasitism studies, the host may be destroyed before identification is possible, and in these situations the species of nonparasitised specimens collected in the same situation should be recorded.

V.   Transporting Live Specimens

Incorrect transport procedures can kill specimens directly, subject them to stress, and make them more susceptible to infections. If lengthy trips are being undertaken, then specimens should be sent back to the laboratory by bus or other means. (Ensure there is someone to collect and process the consignment as soon as it arrives.) Exposure of specimens to sunlight and to lack of air are the situations that are most likely to cause mortality.

For transporting adults, a strong mesh cage with an armhole access is probably the easiest method, but take care that the mesh is not damaged. Crumpled paper in the cage will help keep the moths in good condition, and a source of sucrose solution may be included also if spillages can be prevented. Once adults are in the cage, ensure that the cage is never in direct sunlight, and leave the cage out of the car and in the shade whenever possible.

Eggs and larvae are best collected into containers that have some form of ventilation (for example, cardboard lids or small holes in the lid covered with fine mesh) to reduce condensation. Where cardboard lids are used, records can be kept or the containers numbered (in pencil) on the lid, otherwise information can be written directly onto the containers or onto masking tape (the label type least likely to become detached) stuck on the containers. If parasites are of interest, then clear containers should be used so that parasites emerging during transit can be seen without having to remove the lid of the container.

To prevent cannibalism, adequate food must be provided, particularly where more than one medium to large larva is kept per container. If large and small larvae are kept together, some cannibalism can be expected. Larvae obviously close to pupation should be kept individually, and in containers that prevent them from chewing their way out.

Either host plant material or artificial diet (see Chapter 3) can be used as a food source during transport. Containers should not be stuffed full of host plant material as condensation will be a problem. If artificial diet is used, then it should be poured and set in the collection containers, as blocks of diet cut and placed in collection containers tend to roll around and squash or drown larvae. Immediately prior to use, the surface of artificial diet should be wiped to remove condensation. For eggs and very small larvae, it is best to include only a small quantity of plant material, and not to use artificial diet at all as condensation is a major problem. As soon as they are collected, place specimens in an insulated container. A small foam container is ideal as these are light, and inexpensive to replace. Specimens should be transferred to the larger containers in the vehicle as soon as possible.

For transporting eggs and larvae, insulated containers or a car refrigerator will be a necessity. The latter is the more expensive option, but will maintain cool conditions for extended periods. Because of the ability of these fridges to maintain temperatures below 0°C, care needs to be taken that the collections are not kept too cool (below 15°C) during transport. Foam-insulated containers are fragile and have a limited life, so insulated plastic or metal containers may be best. Whichever type of insulated container is used, it will need ice bricks, and the ice bricks are best wrapped in newspaper or other absorbent material to prevent the condensation that results from specimen containers being stored in direct contact with the ice bricks.

Pupae can be kept in empty unit containers individually, but are less likely to be damaged if tissue or towelling paper is used to cushion the pupae and prevent them from rolling around. Where a number of pupae are kept in one container, extra care needs to be taken to cushion them and prevent them from rolling against one another. Collected pupae should be put into an insulated container immediately, and placed as soon as possible in the larger container in the vehicle.

VI. Laboratory Rearing

After identification, adults can be kept individually, in pairs, or in bulk. In all instances, containers should be well ventilated and have access into the container through a sleeve of material. See Chapter 3 for details of laboratory culturing of *Heliothis*.

It is not uncommon for collected eggs and larvae to be stored for significant periods of time prior to processing. To delay development of specimens, they should be stored at 15°C. If stored below this temperature mortality may be increased, and at temperatures above 18°C appreciable development may occur. In general, storage time should be kept to a minimum.

Condensation is common on removal of egg and larval specimens from cool conditions, so each container should be checked after transport or storage and dried if necessary. During this procedure, specimens could be identified also.

Where eggs are kept singly, multicelled trays covered with plastic film have been successful, particularly when eggs were collected using the Hoffman leaf disc cutter. Providing a small quantity of plant material will ensure that neonate larvae survive. Larvae should be removed to larger containers as soon as possible after hatching. An alternative to trays is clear plastic (to allow observation of parasite emergence) unit containers with ventilation holes covered by very fine mesh (to prevent escape of parasites). Bulk egg collections can be incubated on host plant material or on artificial diet (see Chapter 3) in large containers. Containers should be well ventilated to prevent moisture accumulating. If host plant material is being used as a food source, surface sterilization of material may reduce disease problems.

Young larvae can be kept together in containers if adequate food is provided, but larger larvae (over about 0.5 cm long) need to be kept separately to prevent cannibalism. Again, containers should be well ventilated. If parasitism is being studied, then use clear plastic unit containers and ventilation holes covered with very fine mesh. Prior to pupation, larvae (particularly of *H. armigera*) may chew through container lids; close to this stage, placing containers inside other containers with a layer of pupation medium in the base may help. To pupate and emerge as adults successfully, prepupae should be removed to containers with 2–3 cm of a dry medium such as sawdust (chemical-free) in the base.

VII. Conclusion

Each study undertaken will result in a multitude of data, and the problems of making this information available in a synthesized and comprehensible form will not be overcome until a national register of *Heliothis* collecting records is established.

Chapter 3

Laboratory Culture of *Heliothis* Species and Identification of Disease

R.E. Teakle

I. Introduction

Heliothis spp. are reared in the laboratory for different reasons. These include short-term assessment of samples of field populations for diseases (below), parasitism (Chapters 2 and 7), and resistance (Chapter 12), or for long-term studies requiring the establishment of laboratory cultures for several generations (see Chapters 9, 12, and 13). The approach to the culturing of the insects will differ with the purpose for which they are to serve. For example, a relaxation in standards of food and progeny quality may be permissible if the insects are to be discarded after one or two generations.

Most experience in rearing Australian *Heliothis* species has been obtained with *H. armigera* and *H. punctigera*, and a number of laboratories in Australia maintain cultures. There is some variation in the food composition and methods used (M.P. Zalucki, unpublished). The rearing methods described here have been developed during maintenance of cultures of *H. punctigera* (since 1972) and *H. armigera* (since 1980). The details are the same for the two species, and most of the details have been published (Teakle and Jensen, 1985).

1. Short-Term Maintenance

Larvae can be maintained on natural host plant material, if necessary. Artificial diets are preferred if available, because their use reduces the amount of handling of larvae in the laboratory, and avoids the risk of introducing pathogens on contaminated plant material. *Heliothis armigera* and *H.*

punctigera may be reared to pupation on the relatively simple artificial diet of Shorey and Hale (1965). When the gut contents of larvae collected from the field are excreted onto artificial diet, they may initiate microbial decomposition of the diet. Field-collected larvae should be transferred to fresh diet when the gut contents have been voided, after about 48 h (see Chapter 2). Eggs and larvae can be reared to pupation in 24-well tissue culture trays half-filled with diet and sealed with polypropylene balls (Griffith, Smith, and Williamson, 1979), but this involves stress owing to the barely sufficient space and food supply. Plastic cups (28 ml) ventilated with a cotton wool wick through the lid are preferable if moths or subsequent generations are required.

2. Long-Term Maintenance

Successful rearing of *Heliothis* spp. requires the development of a rearing schedule and a high level of commitment to the task. A daily routine is essential in order to ensure that all aspects of the culturing are attended to.

Newly-hatched larvae should be supplied with food with minimal delay. To reduce the opportunity for larval contact and aggression, diet should be cut into strips so as to provide a large surface area. It is also advisable for operations involving adults to be done early in the morning, when they are not active and are easier to handle.

3. Establishing the Culture

All stages of *Heliothis* from the field may be used to establish laboratory cultures. Using gravid females collected using light traps has the advantage of avoiding the introduction of parasites and most diseases.

Where the insects serve merely as units in, for example, routine bioassays, it would be clearly an advantage to start the colony from an existing culture. Prior selection of the insects for ability to tolerate the handling and "abnormal" conditions within the laboratory would have been made already.

4. Maintaining the Culture

The following factors need to be considered during the maintenance of insect cultures:

1. Food. The larval diet normally comprises a grain legume, wheatgerm, yeast, ascorbic acid, antimicrobial substances, and a gelling agent to give the mixture a firm consistency. The modified Shorey's diet described by Teakle and Jensen (1985) has been used to rear *H. armigera* and *H. punctigera* continuously over 5 years. Diet is prepared as required, poured into flat trays, sealed in polythene bags to retard drying, and stored at 5°C. The diet is normally not used if it is more than 10 days

since its preparation. Moths are supplied with a 10% sucrose or honey solution in a container with a cotton wool wick. We autoclave 10% sucrose at 120°C for 15 min, allow the solution to cool, and add ascorbic acid at 0.5% and streptomycin at 40 mg/l. The streptomycin is used to suppress bacterial diseases.

2. Ventilation. Ventilation becomes increasingly important as larvae develop. Asphyxiation may occur unless loose-fitting lids or lids vented with a cotton wool wick are used. Also, as the larvae approach pupation, it is desirable that the diet dries visibly, as this discourages the growth of mold.

3. Hygiene. This is of paramount importance to prevent highly destructive diseases from becoming established (see below for diagnostic keys). Eggs, pupae, brushes, and working surfaces should be surface-sterilized using sodium hypochlorite (Ignoffo, 1965). Disposable containers should be used and never reused. Forceps should be flamed regularly between handling the insects. It is advisable to have about 10 brushes for transferring small larvae. These are used in rotation and sterilized in 1% sodium hypochlorite. Similarly, about 10 pairs of forceps should be used for larger larvae. The forceps are used in sequence and then sterilized in a Bunsen burner after all have been used. It is desirable to keep the insects in small batches that are isolated from each other, in order to prevent the spread of any pathogen that might breach these defenses. Eggs are surface-sterilized by placing the paper with attached eggs in a photograph-developing tray containing two liters of 0.2% sodium hypochlorite. After occasional agitation for 8 min, the eggs will become detached and settle to the bottom of the tray. Most of the supernatant is poured off and the eggs are recovered by passing the remainder through a Buchner funnel, lined with two layers of absorbent tissue. The eggs are washed with tap water and folded with wet tissues between paper towelling. The eggs must be held in a moist environment, usually in a polythene bag, to avoid desiccation. This is because the sodium hypochlorite treatment removes some of the chorion.

Pupae are surface-sterilized using 0.25% sodium hypochlorite. During the immersion for 10 min, the pupae are held in a plastic sieve, and then washed under running tap water. The pupae are then transferred to a wide-neck clear container lined with paper for adult emergence.

4. Larval aggression. Normally, larvae are reared individually, at least during the later instars, to avoid fighting between the insects. Such aggression can lead to damage or death and possible cannibalism. This is more likely to occur in *H. armigera* than in *H. punctigera*.

5. Moth mating conditions. Moth infertility appears to be a major cause of colony failure. Factors that appear to favor the production of fertile eggs are: adequate cage volume; exposure to natural daylight; temperature between 25°C and 27°C, a high humidity; as well as a supply of a sugar or honey solution. Moths from our established culture produce predominantly fertile eggs. The adults emerge from the pupae directly into cages consisting of aluminium rod frames, 250 mm × 250 mm × 400 mm, fitted with mesh sleeves. However, F_1 or F_2 generation moths from field-

collected insects cannot be relied upon always to produce fertile eggs. Some experimentation, such as the provision of larger cages containing bundles of flowers in water, may be helpful. Moths can be transferred from the mating to oviposition cages using an aspirator (Teakle and Jensen, 1985) attached to a vacuum cleaner. The mating cage should receive back-lighting during the transfer to reduce the risk of moths escaping.

6. Oviposition and egg harvesting. Observation suggests that moths will lay their eggs on a rough, vertical surface. Paper towelling serves as an oviposition site for established cultures. If this fails, natural host flowers may induce oviposition. Eggs laid on paper towelling in cages may be harvested and moths may be retained in the cages by withdrawing the paper through a 3-mm slit, as described in Teakle and Jensen (1985).

   An oviposition cage made of aluminium fitted with a perspex ('Lexan') lid, from a design of P.H. Twine (personal communication), has proved satisfactory (Teakle and Jensen, 1985). The base is perforated to allow moth scales to fall through into a dish of water. The walls consist of paper towel that feeds around a guide and automatically replaces the paper (with adhering eggs) as it is withdrawn. The handler should wear a face mask to reduce the risk of inhaling moth scales, which can be allergenic. If ants pose a problem, then cage supports should stand in paraffin oil.

7. Temperature, day length, and humidity. Colonies can be maintained at a constant temperature of 25°C, or the temperature can be increased to 27°C or 30°C to accelerate the development of eggs or larvae. The period to larval hatch can be adjusted to either 3 or 2 days by maintaining eggs at 25°C or 30°C, respectively (Kay, 1981). To discourage facultative pupal diapause, photoperiod should be at least a 14:10 (light:dark) regimen and temperature above 20°C. The room in which the larvae are reared should be maintained at a relative humidity of less than 65% to discourage mold growth on the artificial diet and to allow the diet to dry as pupation approaches.

II. Daily Schedule for Rearing *Heliothis*

Colonies can be maintained continuously by setting up batches of larvae twice weekly. To avoid the necessity for handling at weekends, newly-hatched larvae are set up on Mondays and Fridays from eggs harvested on Fridays and Tuesdays, respectively. Batches should be as large as practicable with at least 100 larvae, in order to maintain a large gene pool. We select half of the larvae that are the largest at 7 days, and discard the remainder. We rear all stages at 25°C. Our schedule is as follows:

Monday. Eggs laid during the weekend are harvested. Newly-hatched larvae are transferred to diet, and maintained in disposable containers in batches of about 10 larvae. Seven-day-old larvae are transferred to fresh diet in individual containers, normally 28-ml plastic cups with waxed cardboard tab lids (Lily, P100). These allow adequate ventilation for the larva and partial drying of the frass in which the larva pupates.

Tuesday. Eggs are harvested into plastic bags and held at 25°C.

Wednesday. Eggs may be harvested and held at 30°C if additional newly hatched larvae are required on the Friday.

Thursday. Eggs are cleared from oviposition cages and those eggs that were harvested on Tuesday are surface-sterilized. Pupae derived from larvae set up a month earlier are harvested and surface-sterilized.

Friday. Eggs are harvested and surface-sterilized. Newly-hatched larvae are set up in batches of about 10. Seven-day-old larvae are transferred to individual containers.

The following should be checked daily (Monday to Friday) and transfers made when appropriate:

(1) Pupae are transferred to mating cages where adults emerge;
(2) Adults in mating cages are transferred to oviposition cages when eggs are evident on the walls of the mating cages;
(3) Adults in oviposition cages are destroyed by autoclaving after egg production drops to a low level, usually in about a week.

III. Diagnosis of Diseases in *Heliothis*

The diagnosis and recording of diseases in *Heliothis* populations is required to understand their prevalence and role in the ecology of these insects (see Chapter 6). Early diagnosis is also essential in maintaining healthy laboratory cultures. At least 10 pathogens associated with *Heliothis* spp. have been recorded (Ignoffo, 1971). Poinar (1975) has described a nematode with an associated bacterium in *H. punctigera* pupae from South Australia. Diseases recorded in Queensland with photographs of infected *Heliothis* spp. larvae, have been published (Teakle, 1977). Details of the diseases are given briefly in Table 3.1.

1. Viruses

A. *Nuclear Polyhedrosis Virus*

Nuclear polyhedrosis is a commonly recorded cause of epizootics in *Heliothis* spp. populations. The virus has been developed as a commercial insecticide (Ignoffo, 1973). A full description of the disease in *H. punctigera* has been given (Teakle, 1973). Infected insects display little or no signs of the disease until 1 to 2 days before death, when a creamy pallor and shiny appearance due to distension of their hypodermis may be seen. Dissection reveals that the organs principally infected, the hypodermis, fat body, and tracheal matrix, have a spotted appearance due to the packing of many cells with the virus polyhedra. The infected larvae tend to move to the top of the

Table 3.1. Key for Pathogens of *Heliothis* Larvae

1.	Dead larva mummified with external hyphae	2
	Dead larva, soft or dried, usually lying upon or hanging from plant surface	3
2.	Conidia green, oval, in chains on the end of phialides	*Nomuraea rileyi*
	Conidia white, spherical, attached to alternate sides of conidiophores	*Beauveria bassiana*
3.	Skin fragile, body contents liquified and containing masses of polyhedra approx. 1 μm in diameter	Nuclear polyhedrosis virus
	Body contains large number of small, usually motile, rod-shaped cells or oval spores approx. 1 $\times$ 1.5 μm	Bacteria
	Skin intact, creamy pallor	4
4.	Masses of minute particles in suspension of fat body viewed at high magnification	Granulosis virus
	Large ovoidal spores approx. 5.5 $\times$ 3.5 μm in suspension of body contents	*Nosema heliothidis*

plant, spin a sparse silken mat, and become attached by the prolegs. At death, the larvae are soft, fragile due to the breakdown of the hypodermis and other tissues, and usually rupture releasing a sticky, virus-laden fluid. After death, the insect rapidly darkens and may be invaded by secondary microorganisms. A suspension of the body contents examined microscopically at about $\times 250$ under phase contrast shows masses of refractile crystals approximately 1 μm in diameter.

Usually, the virus is acquired by ingestion of contaminated food or by cannibalism of infected larvae. Egg contamination can occur, in which case emerging larvae usually die in the first instar. Small larvae are the most susceptible and die within 3–5 days of exposure to the virus. Larger larvae usually die within 4–7 days of infection, but may succumb after as long as 2 weeks, or survive large doses. The infection of newly-hatched larvae could theoretically result in several cycles of virus release within the one generation (R. Teakle, unpublished data). The rapid breakdown and release of infectious polyhedra allows rapid spread of the virus in dense populations of larvae.

B. Granulosis Virus

The incidence of granulosis may equal or exceed that of nuclear polyhedrosis in low density populations of *Heliothis* spp.. In high density populations, the nuclear polyhedrosis usually predominates.

After ingestion of the virus, the infection becomes localized in the fat body. Infected larvae always die in the fourth or later instars (Whitlock, 1974). Larvae infected in the later instars die in the final instar as large, swollen, creamy-yellow larvae within 2–3 weeks, by which time normal larvae have pupated. Apparently, the hypodermis is not infected and the larvae remain intact at death (Teakle, 1974). However, bacteria present occasionally cause early breakdown of the cadaver. Examination of a suspension of the infected fat body under phase contrast at a magnification of approximately ×900, reveals masses of minute virus particles that settle out on the microscope slide as a dark stippling.

The prolonged incubation periods and delayed release of the virus probably account for the relatively low incidence of the virus in high larval populations. Granulosis and nuclear polyhedrosis infections can occur within the same larva. In this event, the larger virus polyhedra usually mask the presence of the granulosis virus in suspensions of the body contents. Laboratory evidence suggests that the granulosis virus will inhibit infection by the nuclear polyhedrosis virus unless an advantage in time or concentration is given to the nuclear polyhedrosis virus (R. Teakle, unpublished data).

2. Bacteria

Frequently, bacteria have been recorded in field-collected larvae that have died while being maintained in the laboratory. The pathogenic status of the bacteria generally has not been determined. At death, larvae often show a pinkish discoloration. The body then darkens to brown or black. The skin usually remains intact and a faint disagreeable odor may be present. Bacterial pathogens may be spore-formers or non-spore-formers, and are of two broad types: (1) primary pathogens that possess some mechanism for harming the insects; and (2) potential pathogens that cause septicemia after prior injury to the insect (Bucher, 1960). Bacteria present may be nonpathogenic and involved in the decomposition of the dead insect.

Bacillus thuringiensis has been isolated from dead *Heliothis* larvae in our laboratory culture and apparently originated from Lima beans used at the time in the artificial diet. This bacterium appears as large, rod-shaped cells that produce oval spores and toxic diamond-shaped crystals after several days' growth on nutrient agar. A similar rod-shaped bacterium, *Bacillus cereus* that produces spores, but not crystals, has also been noted. Masses of spores in the dead insect are easily mistaken for polyhedra from nuclear polyhedrosis virus disease.

Such small, non-spore-forming rod-shaped bacteria as the blue-green pigmented *Pseudomonas aeruginosa* and the red-pigmented *Serratia marcescens*, have been isolated from dead laboratory-reared *H. armigera*. Strains of these bacteria have the capacity to cause septicemia (Steinhaus, 1959; Lysenko, 1963; Bell, 1969).

Non-spore-forming bacteria can be distinguished readily from nuclear polyhedrosis virus, in that the bacterial cells appear dark under phase contrast and are frequently motile, whereas the virus polyhedra are pearly bright and tend to settle onto the microscope slide.

3. Fungi

Nomuraea (Spicaria) rileyi commonly infects *Heliothis* spp. and other noctuid larvae on a range of crops, usually those with abundant foliage. Occasionally, epizootics are seen. The potential for the fungus as a microbial insecticide has been studied in the United States (Sprenkel and Brooks, 1975; Ignoffo, Marston, Hotstetter, Puttler, and Bell, 1976).

Field-collected larvae often succumb to the disease in the laboratory. Early signs are stiffening of the insect and its attachment to the substrate. White hyphae emerge at the intersegmental (and other soft) areas and eventually envelop the larva. Under dry conditions, development of the fungus may be arrested at these early stages. With high humidity, conidia form and the larva becomes enveloped in a blue-green velvety covering. Microscopic examination of a small piece of this covering reveals that the oval conidia are borne in chains on the ends of phialides (Weiser and Briggs, 1971). The fungus can be isolated and cultured on artificial media (Getzin, 1961; Bell, 1975).

Beauveria bassiana has a wide host range and occasionally has been recorded on *Heliothis* spp. larvae. Fungal development proceeds similarly to that of *N. rileyi*. After conidia formation, the larvae are enveloped in a white mass resembling cotton wool. The spores are spherical and typically occur attached to alternate sides of the conidiophores (Weiser and Briggs, 1971).

4. Microsporida

The microsporidian, *Nosema heliothidis* has been recorded infrequently in *Heliothis* spp. in Australia, although it is commonly associated with those *Heliothis* species found in the United States (Brooks and Cranford, 1975). Infections are usually sublethal and effects on populations tend to be debilitative rather than destructive (Gaugler and Brooks, 1975). The disease probably often goes unnoticed. Larvae killed by the disease are similar in external appearance to those with granulosis, that is, creamy-yellow and skin intact. Microscopic examination of the body contents will reveal the presence of large ovoid spores measuring 2.85–6.55 μm in length and 1.25–3.5 μm diameter (Kramer, 1959; Brooks, 1968). Larvae can be infected per os. Infected adults may transmit the infection transovarially (Brooks, 1968).

Chapter 4

Trapping Methods for Adults

P.C. Gregg and A.G.L. Wilson

I. Uses of Traps

Light and pheromone traps are important tools in ecological research and pest management with *Heliothis* spp. Population surveys may be undertaken for the purposes of

1. determining the presence or absence of *Heliothis* spp. in an area;
2. collecting specimens for taxonomic purposes, or for establishing laboratory cultures;
3. determining the relative proportions of *H. armigera* and *H. punctigera* to assist in choice of insecticides;
4. detecting migration;
5. providing early warnings of crop infestations and oviposition;
6. obtaining quantitative estimates of population density, species composition, or age and sex structure for studies of distribution, phenology and population dynamics.

Traps may also be used as a method of population suppression in pest management. Attempts to do this with *Heliothis* spp. have been largely unsuccessful, and this use of traps is discussed only briefly here.

An ideal trap should be cheap, durable, and robust. It should be serviced easily by personnel of little training, should be highly efficient, attract a large number of moths, and retain as many of them as possible. The trap catch should reflect the population of the area in total numbers, species composition, sex ratio, and mated status. Unfortunately, such an ideal trap does not exist. Here, we review the traps that are available, their advantages and disadvantages in the roles listed above, and some general problems associated with trapping.

1. Presence/Absence Studies

In many areas of Australia, the presence of *Heliothis* spp. is well known (Zalucki et al., 1986). There are some areas where lack of distribution records may reflect lack of cultivated hosts and of collecting effort rather than the absence of the species. Gregg, McDonald, and Bryceson (1989) used traps to establish the presence of *Heliothis* spp. in the arid south-west of Queensland, some 500 km further west than previous records.

For such purposes, the most important features of a trap are ease of use in remote areas and the ability to trap moths from low density populations efficiently. Accurate reflection of population density, species composition, and age and sex structures are less important because only a few captures are needed to establish presence. Good preservation of the specimens is not needed so long as identification is possible.

2. Collecting

The use of light traps is a standard method for collecting specimens to be pinned or dissected in taxonomic studies. In this case, good preservation of the specimens is essential and quick killing agents are required. Devices to segregate moths from such insects as scarab beetles, which might abrade them by trampling in the trap, may be needed. If the moths are to be trapped live, for example, to establish a laboratory culture, a large holding container in which moths can survive even in extreme weather conditions is essential.

There is little need for collecting to reflect population density, species composition or age and sex structure accurately. Pheromone traps that catch only males are of limited value.

3. Determining Species Composition

If the species composition of adults caught in traps can be related to that of the eggs laid in crops, then it assists in the selection of insecticides, particularly when pyrethroid-resistant *H. armigera* may be present. Electrophoresis of the eggs and small larvae can provide this information directly (Daly and Gregg, 1985), but traps are easier to operate at the farm level and might also give an earlier warning of changed species composition. The critical requirement for traps in this role is that species composition in the trap should be correlated closely with that of eggs laid in the field.

4. Detecting Migration

Traps can be used to detect moths in an area before local emergence occurs, thereby providing indirect evidence of migration. Hartstack et al., (1986) used pheromone traps in this way to detect migration of *H. zea* into Texas and the Mississippi valley from areas to the south. Comparisons of light trap catches and local emergence of *H. punctigera* at Narrabri also suggest migration (Wilson, 1983). When used for these purposes, the qualities required of a trap are similar to those for presence/absence studies. The main difference is the requirement for more frequent trap-servicing in order to differentiate population peaks of immigrants from local moths.

Another use of traps in migration studies is their placement in areas where all moths caught should be migrants (see Chapter 10). Sparks, Jackson, and Allen (1975) captured *H. zea* in light traps on oil platforms up to 160 km offshore in the Gulf of Mexico, thereby establishing the capacity of that species to fly at least that distance. Callahan, Sparks, Snow, and Copeland (1972) used upward-facing light traps mounted on a television tower in Georgia to demonstrate that *H. zea* could be collected at altitudes of up to 320 m, at which height long-distance migration probably occurs.

5. Predicting Oviposition

Trap catches are used for detecting initial infestations and subsequent increases in activity in many crop–pest interactions. In some cases, insecticidal or other control methods can be timed at least partly on the basis of trap catches. Examples are provided by *H. zea* in Texas corn (light traps, Lopez, Hartstack, Witz, and Hollingsworth, 1979) and *H. virescens* in South Carolina cotton (pheromone traps, Johnson, 1983). In these cases, trap catches accurately reflected oviposition in the crops and this is the crucial factor in determining the value of traps in this role. In Australia, hopes that pheromone trap catches could be used to time insecticide applications (Rothschild, Wilsons, and Malafant, 1982) have not been realized yet. Data from both light and pheromone traps have been available to growers as an indication of general activity from the SIRATAC database (see Chapter 14), but accurate predictions are not possible yet.

6. Quantitative Studies

Providing data for quantitative studies of distribution, phenology, and population dynamics is the most difficult role for trapping. The requirements are similar to those for determinations of species composition and oviposition patterns, but the accuracy needed is much greater. For the purposes of pest management, it may be sufficient to categorize activity as high, medium, or low. For quantitative ecological work, the greater the precision, the better. The many factors affecting the correlation between numbers, species composition, sex ratios, and mated status in traps with the corresponding parameters in the field are discussed later in this chapter.

Despite the difficulties, some overseas workers have attempted to use trap data in population dynamics studies or simulation models (Hartstack, Hollingsworth, Ridgeway, and Coppedge, 1973; Hartstack and Witz, 1981). In

Australia, such work has been limited to approximate timing of generation intervals from light trap data (Wardhaugh, Room, and Greenup, 1980; Morton, Tuart, and Wardhaugh, 1981; Wilson, 1983).

7. Population Suppression

Traps can lower the growth potential of a population by removing adults. Light traps have been used to suppress the tobacco hornworm *Manduca sexta*. Satisfactory control was achieved by a density of three traps per square mile (Cantello, Smith, Baumhover, Stanley, and Henneberry, 1972). The major requirements for traps used in this way are the capacity to attract moths from a long distance, trap them efficiently, and minimize servicing needs.

Hartstack, Hollingsworth, Ridgeway, and Hunt (1971) calculated that the minimum spacing required for control of *H. virescens* with light traps would be about 100 to 200 m, suggesting that the method might have applications, but probably only for small areas of high value crops. This conclusion is supported by the results of Sparks, Wright, and Hollingsworth (1967), who tested several large commercial light traps for their effectiveness in controlling *H. zea*. Even though catches were high at a spacing of 170 m, when the spacing of the traps was halved the total catch was almost doubled, indicating that the original spacing left many moths untrapped. Sparks, Raulston, Lingren, Carpenter, Klun, and Mullinx (1982) discussed the use of pheromone traps to suppress populations, but concluded that the limited efficiency of the traps and the fact that peak catches of males occurred after mating had taken place indicated that few possibilities exist.

II. Trap Design

1. Light Traps

Light traps used to catch *Heliothis* spp. vary in design but generally consist of three components: (1) a light source; (2) an arrangement of baffles around the source; (3) and a catch container with a killing agent.

A. Light Source

Positively phototropic insects are attracted to the contrast between a light source and its surroundings, though high light intensities in the immediate vicinity of the source may repel them (Verheijen, 1960; Bowden, 1982). Different light sources vary in their output, intensity, spectral distribution, shape, and size.

As the output of radiant energy of a light rises, so does the catch of the trap. However, the increase is not proportional. Bowden (1982) analyzed the results of a number of studies and showed that the catch rose only as the square root of the output, thus accounting for the many observations of surprisingly good performance by small light traps (Smith, Taylor, and Apple, 1959; Belton and Kempster, 1963). As well as the total output, intensity (output per unit area) of the source might be important, as it determines the contrast between the trap and its surroundings. For some light sources, intensity increases with output. An exception is the fluorescent tube, in which increased output is gained by increasing the size of the tube rather than the intensity.

Often some compromise must be made between trap attractiveness and power consumption. Mercury vapor lamps are limited to sites where grid power is available or can be supplied by a generator. Small fluorescent tubes can be run by a 12-V battery. Battery life may vary from one night in the case of 40-W tubes to a week or more in the case of 6-W tubes.

In regard to spectral composition, most workers agree that the shorter visible and near ultraviolet (UV) wavelengths of between 320 and 600 nm are most attractive to insects in general (Dethier, 1963; Southwood, 1978), but there is considerable variation between species. Even within the genus *Heliothis* there may be variation. Hendricks, Lingren, and Hollingsworth (1975) found that catches of *H. zea* were higher in a trap containing a blacklight (BL) tube with an emission peak in the ultraviolet range at 365 nm than were those in one containing a green tube with an emission peak at about 525 nm. For *H. virescens,* the green tube was the most effective. Electrophysiological studies of the eyes of the two species give no clues to the mechanisms of such differences, as both species showed almost identical peaks of responsiveness in the two regions (Agee, 1973).

The size and shape of the light source have received little attention, probably because in commercially available sources they are confounded with differences in such variables as intensity and spectral composition. Bowden (1982) provides diagrams showing the spatial distribution of light around some common sources. There is substantial variation between apparently similar types of mercury vapor bulbs, and major differences between bulbs and fluorescent tubes. The latter have much higher intensities in the horizontal plane than at angles above or below. Therefore, they might attract fewer high-flying insects than would traps with a bulb as a light source, but critical studies are lacking. It probably matters whether tubes are oriented horizontally or in the conventional vertical position. Horizontal UV tubes caught 2.6 times as many house flies as did vertical tubes in illuminated rooms (Pickens and Thimijan, 1986). Critical tests with *Heliothis* are lacking.

In Australia, four types of lights are commonly used in traps for *Heliothis* spp., that is, two mercury vapor bulbs and two UV-emitting fluorescent tubes. The spectral composition of their lights are illustrated in Figure 4.1.

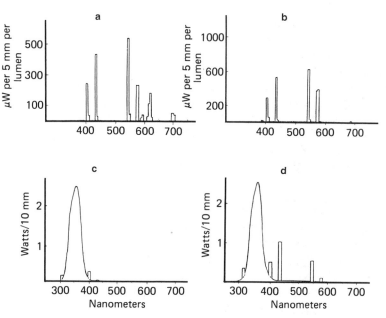

Figure 4.1. Spectral distribution of light sources used in Australian light traps for *Heliothis* spp. The mercury vapor bulbs have an additional emission band in the ultraviolet at about 365 nm, which is not shown in these diagrams: a, true mercury vapor bulb (Philips HPL-N-125W) (drawn by P.C. Gregg with permission from Philips Lighting Industries); b, blended mercury vapor bulb (Philips ML-160W) (drawn by P.C. Gregg with permission from Philips Lighting Industries); c, blacklight blue fluorescent tube (Sylvania F40 BLB) (drawn by P.C. Gregg with permission from Sylvania Electric Australia); d, blacklight fluorescent tube (Sylvania F40 BL) (drawn by P.C. Gregg with permission from Sylvania Electric Australia).

True mercury vapor bulbs, such as Philips HPL-N-125W, have line spectra with main peaks at 5 wavelengths. Blended mercury vapor lamps such, as Philips ML-160W, contain a tungsten filament as well as a mercury vapor discharge tube. Thus, both types emit substantial amounts of both visible and UV light, but the proportion of UV is slightly lower in the blended type. Both types give a light that is bright bluish-white to the human eye. Taylor and Brown (1972) compared the catches obtained with both types in Kenya, and found that the blended type caught slightly more than half the number of Lepidoptera than did the mercury vapor type. Mercury vapor bulbs have a much higher intensity than do blended types, but there may be substantial variation between individual bulbs of both types (Bowden, 1982). The mercury vapor bulb requires a ballast, whereas the blended bulb does not. Both types become very hot during operation and must be covered to prevent shattering if rain falls on them. Usually this is accomplished by a transparent heat-resistant beaker or glass jar, but the effect of this on the transmission of UV light should be considered. Generally, mercury vapor bulbs are fitted to traps of the Robinson type (Robinson and Robinson, 1950). An example is the light trap at the Agricultural Research Station, Narrabri, Australia (Fig. 4.2a).

Fluorescent tubes that emit UV light are the blacklight (BL) type, sometimes called actinic blue, and the blacklight blue (BLB) type. Both have major peaks in the UV region at around 365 nm, but the BL type also has peaks in the visible range around 430 nm and 540 nm (Fig. 4.1). To the human eye, the BL type gives a bright bluish-white light, but the BLB type gives only a dull blue glow that is difficult to see from more than 100 m or so.

L.G. Boyce and P.C. Gregg (unpublished) have compared the catches of traps equipped with 6-W versions of these tubes with those using ML-160W (Philips) bulbs. The catch of all insects with BL tubes was 82% of that with the ML-160W bulbs, whereas that of the BLB tubes was 39%. The composition of the catch with BL tubes was similar to that with the ML-160W bulbs, but there were significantly fewer moths and more beetles with the BLB tube. This is difficult to reconcile with the comments of Southwood (1978) that moths are attracted to UV light, but it is in agreement with the experience of Commonwealth Scientific and Industrial Research Organisation (C.S.I.R.O.) lepidopterists, who prefer the BL tube over the BLB (E.S. Nielsen, personal communication).

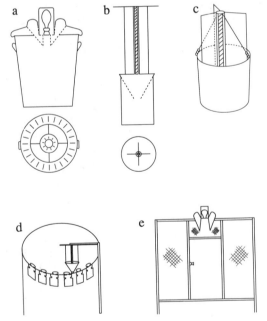

Figure 4.2. Light traps: a, Mercury vapor Robinson trap based on garbage can; b, Pennsylvania trap with 40-W UV tube; c, CSIRO collecting trap with 8-W UV tube; d, Rotating carousel trap with 6-W UV tube; e, Robinson trap mounted on walk-in cage for live collection (all drawn by P.C. Gregg).

Fluorescent tubes are fitted to traps of the Pennsylvania type usually (Frost, 1957). Large traps using 40-W BLB tubes are used by the Agricultural Research Station at Narrabri (Fig. 4.2b), whereas smaller ones with 8-W BL tubes are used by lepidopterists at the CSIRO Division of Entomology, Canberra, Australia. (Fig. 4.2c; Olsen, Nielsen, and Skule, 1984). Still smaller Pennsylvania traps with 6-W BLB tubes are fitted to the automatic traps shown in Figure 4.2d. The power consumption of the latter is so low that a solar charger is able to recharge the battery after each night and the trap can operate indefinitely without grid power or battery replacement.

B. Baffles

Most light traps are fitted with baffles, often four arranged perpendicularly to each other. The object is to retain those insects that would otherwise have missed the trap, perhaps because of their avoidance reaction close to the light source or perhaps because of turbulent airflow around the trap. Generally, it is believed that baffles greatly increase the catch of a trap, but there are few quantitative studies. Frost (1958) found that the addition of baffles to a 15-W BL tube increased the catch by about 82%. Most of the increase was in the smaller Diptera and Microlepidoptera. The catch of noctuid moths was increased by only about 27%.

The design of baffles is influenced by the type of light source. Traps with bulbs, such as Robinson traps, usually have broad, low baffles, whereas those with tubes, such as Pennsylvania traps, have high, narrow ones (Figs. 4.2a–d). Baffles may be made from an opaque material such as galvanized iron, or from a transparent material such as clear acrylic. It is possible that the transparency or other characteristics of the baffles affects the insect's reaction close to the trap (Robinson, 1952). Certainly, it will affect the spatial distribution of the light. There are few studies on the influence of baffle transparency, although Common (1959) designed a trap that was made entirely of clear acrylic. He reported that the amount of moths caught was similar to that of an opaque trap, whereas that of beetles, especially scarabs, was lower.

C. Containers

In most traps, insects are removed from the light source or baffles by falling into a funnel, from where they are transferred to a killing and holding container. The choice of containers and killing agents is dependent on the objectives of trapping. A large container is desirable where good condition of the specimens is a critical requirement, as in taxonomic studies. It minimizes the damage that results from fluttering inside the container and from trampling by other insects, particularly large hard-bodied ones such as scarab beetles. For live trapping it may help if the trap is mounted above a walk-in cage as in the trap at the Tamworth Agricultural Research Center, Australia (Fig. 4.2e). On the other hand, a small container has the advantage of compactness and ease of transportation, which may be critical for work in remote areas. If several containers are needed on the same trap as in automatic segregation traps (Fig. 4.2d), compactness is essential.

There are a number of container designs that segregate the large insects from the small ones, or those that can fly from those that cannot (Common, 1959; White, 1964). The advantage of these designs is that they minimize damage by trampling to *Heliothis* spp. The same result can be achieved more conveniently but less efficiently by placing papier-mâché egg cartons or crumpled newspapers in the container.

D. *Killing Agents*

The ideal killing agent is fast-acting, cheap, nonrepellent, nontoxic to humans, and longlasting. None of the available ones meet all these requirements. Dichlorvos is commonly used and can be obtained readily in such resin-impregnated strips as "Sheltox"[1] used for the control of household pests. The strips last for up to 2 months in most conditions. The cost can be reduced by impregnating dental rolls or plaster blocks with dichlorvos. Dichlorvos is a slow killing agent, particularly when used on large beetles, and its use often leads to poor quality specimens. A faster killing agent is 1,1,2,2-tetrachlorethane, which is commonly used for taxonomic studies (Olsen et al., 1984). Unfortunately, it is highly volatile, requiring replenishment each night and is quite toxic to humans.

An alternative approach is the use of wet preservation, involving 70% ethanol or methanol as both a killing and preserving agent. This method kills quickly and yields specimens of good quality for identification on the basis of wing markings and for dissection. It does not give specimens suitable for pinning because of the distortion of the wings during drying. The main problem is the high volatility of ethanol, which must be replenished daily under warm conditions. Attempts to reduce this problem by the addition of mineral or vegetable oil to the ethanol have been unsuccessful, as those substances that effectively reduce evaporation also discolor the specimens (P.H. Twine and P.C. Gregg, unpublished). A reduction in evaporation can be achieved by the addition of 10% of ethylene glycol to the ethanol. This extends the life of the killing agent to several days (P.C. Gregg, unpublished), but is expensive.

E. *Other Design Features*

Automatic light traps can segregate the catch into different nights or different periods within nights without the intervention of the operator (Williams,

---

[1]Shelltox® is a registered trademark of Shell Chemical (Australia) Pty. Ltd.

1935). These lights involve some type of rotating carousel that positions a new catch container under the light at predetermined intervals. An example which is being used in studies on Australian *Heliothis* spp. is shown in Figure 4.2d (P.H. Twine, P.C. Gregg and G.P. Fitt, unpublished). It has 14 containers mounted on a circular carousel that moves one place at sunset each evening. Thus, catches can be stored separately between service intervals for up to 14 nights. Such traps provide more data for less manpower, but they are demanding in their requirements of electronic design, killing agents, and holding containers.

Another useful feature is an automatic on/off switch. These may be of the light-sensitive or timer type. The former are simple but susceptible to repeated on and off switching if light intensity varies about the critical level, as happens when there is a low cloud bank partially obscuring the sun at sunset. This may be a particular problem with automatic traps in which the rotation of the carousel is triggered by the on/off switch. A switch recently designed by CSIRO Division of Entomology (B. Condon, personal communication) overcomes this problem by requiring a prolonged drop in light intensity before it is activated. Time switches may be used for on/off switching. The programmable type is accurate, has a time resolution of 1 min, can be programmed for up to 8 on/off switches in 24 h, and can be converted to 12-V operation readily.

2. Pheromone traps

Pheromone traps consist of a lure containing a sex-attractant pheromone for male *H. armigera* or *H. punctigera*, a dispenser for releasing the pheromone, and a device for trapping and retaining the moths attracted.

A. Lures

As with many noctuid moths, males of *Heliothis* spp. are attracted to a mixture of chemicals that must be present in the correct proportions. There has been a continued improvement in the formulations for both species. The currently used blend for *H. punctigera* is 50:50:1 of (Z)-11 hexadecenal, (Z)-11 hexadecenyl acetate, and (Z)-9-tetradecenal (Rothschild et al., 1982). For *H. armigera*, the blend is 10:1 of (Z)-11 hexadecenal and (Z)-9 hexadecenal. This is somewhat different to the blend favored for *H. armigera* in Asia, that is 97:3 of (Z)-11 to (Z)-9 hexadecenal, but produces higher catches in Australia. Different components of the blends diffuse from lures at different rates, despite the addition of antioxidants and substances to reduce volatility (Wilson, 1984). The effects of this on attractiveness of the lures are not fully understood.

B. Dispensers

The purpose of a dispenser is to release an even concentration of pheromone into the atmosphere over as long a period as possible. Excessive release rates early in the life of the dispenser may result in such high concentrations that they depress catches (Gothilf, Kehat, Dunkleblum, and Jacobson, 1979). Later the levels may be too low to be attractive, thus necessitating frequent lure changes. Lures are now becoming available that allow a precise rate of release of pheromone over an exact time (16 weeks for *Heliothis* spp.). However, these lures had not been tested under Australian conditions at the time of writing.

Many types of dispensers have been tried. They include cigarette filters (with or without glass tubes around them), dental rolls, thin hollow fibers, rubber tubing or stoppers, and laminated plastic. In Australia, Wilson (1984) compared four types, that is, a 1.5 cm length of surgical tubing impregnated with pheromone, and commercially produced rubber septa, hollow fibers and plastic laminates. He found that up to 4 weeks exposure at daily temperatures of 22 to 36°C did not reduce the catch of either *H. punctigera* or *H. armigera* with the septa and laminate dispensers, but there were substantial reductions after only 2 weeks with the rubber tubing and hollow fibers. Work in the United States also demonstrated the effectiveness of the plastic laminate (Hendricks, Hartstack, and Shaver, 1977) and it is now the standard dispenser in Australia. Lures for both species are manufactured by Hercon Ltd. in the USA.

C. Traps

Traps used in conjunction with pheromones fall into two basic types, those which moths blunder into in their approach to the pheromone (electric grid traps, sticky traps, water traps, dry funnel traps, etc.) and those which exploit the behavioural response of the moths when they have landed after following a pheromone plume (cone traps and wind-oriented vane traps). The latter are usually much more efficient at capturing moths than are the former.

Electric grid traps consist of a grid of wires surrounding the lure in which alternate wires bear alternate electrical charges of sufficient magnitude to kill or stun moths that contact two wires (Fig. 4.3a). Such traps were widely used in early work on *H. zea* and *H. virescens* in the USA (Mitchell et al., 1972; Goodenough and Snow, 1973; Hendricks et al., 1977). They may be quite efficient, capturing up to 34% of responding moths (Goodenough and Snow, 1973; Lingren, Sparks, Raulston, and Wolf, 1978b), but they have not been adopted in Australia for a number of reasons. The required potential is several thousand volts, and although it is possible to obtain this from a battery, the circuits needed are complex and costly. Electric grid traps are restricted essentially to sites where grid power is available. Also, the voltage needed to keep the grid clean may be so high that specimens are burned.

Sticky traps consist of one or more surfaces close to the lure that are

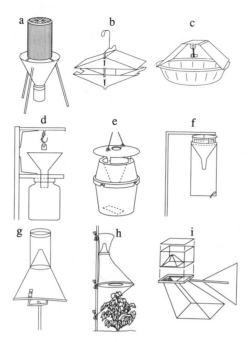

Figure 4.3. Pheromone traps (not to scale): a, electric grid trap (Haile et al., 1973, drawn by P.C. Gregg with permission from American Entomological Society); b, sticky trap, wing type (drawn by P.C. Gregg); c, water trap (Bacon, Sieber, and Kennedy, 1976; drawn by P.C. Gregg with permission from American Entomological Society); d, Australian dry funnel trap (drawn by P.C. Gregg), e, International Pheromones dry funnel trap (drawn by P.C. Gregg with permission from International Pheromone Systems Ltd); f, ICRISAT dry funnel trap (drawn by P.C. Gregg); g, cone trap (Hartstack, Witz, and Buck, 1979a; drawn by P.C. Gregg with permission from American Entomological Society); h, commercial Texas trap, showing correct adjustment of height to crop canopy (Albany International, redrawn by P.C. Gregg); i, Brownsville trap (from Hartstack et al., 1979b; drawn by P.C. Gregg with permission from A.W. Hartstack).

coated with a substance such as Tanglefoot®[2]. Moths fluttering around the lure become trapped in this material. Various designs for *Heliothis* spp. have been described by Hendricks et al., (1972, 1973a), Goodenough and Snow (1973), and Rothschild (1978). Another design that has been adapted for use with *Heliothis* spp. is the wing trap used for monitoring codling moths (Sage and Gregg, 1985; Fig. 4.3b). The main problem with sticky traps for *Heliothis* spp. is that the moth is larger than other species for which the design is successful and the numbers caught are often higher. Consequently, the

---

[2]Tanglefoot® is a registered trademark of The Tanglefoot Company, Grand Rapids, Michigan, 49504 USA.

sticky surface soon becomes coated with moths and scales, and the efficiency of the trap declines rapidly (Wilson, 1984; Sage and Gregg, 1985).

Water traps have a container of water, often with detergent added, in which the moths are drowned (Fig. 4.3c). They have been used with pheromone lures for a number of species, but rarely with *Heliothis* and then only in preliminary studies (Rothschild et al., 1982). They are inconvenient to maintain and not particularly efficient.

The dry funnel or canister trap (Fig. 4.3d) was developed by Kehat and Greenberg (1978). It consists of a funnel with the lure suspended above it and a container with a dry killing agent beneath. A version widely used in Australia uses a 17.5-cm powder funnel and a plastic dinner plate as a roof. Drain holes are provided in the bottom of the container. The killing agent is dichlorvos. The distance between the roof and the top of the funnel has been reduced to 5 cm because this gives increased catches (A.G.L. Wilson, unpublished). The dry funnel trap is not very efficient, and its catch may be reduced by rainfall (Sage and Gregg, 1985). Also, the catch is often biased towards *H. armigera* more than is the case with other designs (Wilson and Morton, 1989). However, it does have the advantage of cheapness and rugged design and most versions are home made. Despite these problems, the dry funnel trap is presently the Australian standard. A compact one-piece plastic version is made by International Pheromones in the UK (Fig. 4.3e). This trap is being used successfully in work done in the USA. (B. Gregory, personal communication). Another variant has been adapted by International Crops Research Institute for the Semi-Arid Tropics (ICRISAT) for Indian conditions (Fig. 4.3f). It is constructed with an aluminium pie dish roof and a 21-cm funnel; moths are collected in a clear plastic bag.

A further modification includes a 15-cm perforated funnel that is inverted and fixed to the roof. It encloses the lure and tends to knock any moths impacting it down into the funnel. This modification produces higher catches, but has not been adopted widely. Studies with night vision devices indicated a maximum catch efficiency of 16% at a wind speed of 0.8 m/s for this trap (D.R. Dent, personal communication).

Cone traps (Fig. 4.3g) were designed by Lingren et al. (1978b) and Hollingsworth, Hartstack, Buck, and Hendricks (1978) following nocturnal observation of the behavior of alighting male moths that tend to climb vertical objects, perhaps as a response to moonlight. Inverted cones that have a retaining cage at the top are therefore effective traps. Hartstack, Witz, and Buck (1979a) compared several designs made of hardware cloth and established the importance of a skirt around the lower rim in preventing escapes. Trap efficiency rose from 5 to 6% to 25 to 30% when this feature was included.

The cone trap with a skirt is known as the Texas trap (Fig. 4.3g). Different types are described by the diameters of the outer rim and the inner skirt, for example the 75:50 Texas trap. Commercial versions are made by Albany

International Corp. in the USA (Fig. 4.3h). These are collapsible and made of a white plastic mesh. They are highly efficient, producing catches of 2 to 10 times those of dry funnel traps (Wilson, 1984; Sage and Gregg, 1985). At Narrabri, locally made versions of the 75:50 Texas trap caught three times the number of *H. armigera* caught by the ICRISAT trap and six times that caught by the Australian dry funnel trap. The catch of *H. punctigera* was eight times that of the Australian dry funnel trap, indicating that the bias towards *H. armigera* may be less severe in Texas traps than in dry funnel traps (Wilson and Morton, 1989). Texas traps have some disadvantages. The plastic mesh in commercial versions deteriorates rapidly in Australian conditions. This appears to be a problem in the United States, as most versions used there are now made of the original rigid mesh or hardware cloth. Wilson and Morton (1989) have made a version using a heat-and-light resistant shade cloth. Another difficulty with Texas traps is that the retaining cage at the top is of open mesh, making it impossible to use a killing agent. The condition of the moths may be poor because of fluttering, and removal of live moths may be difficult. Attempts to modify the Texas trap by fitting a solid detachable container were not successful (Sage and Gregg, 1985), perhaps because the container was opaque. Current U.S. cone traps have a detachable top container, but it is still of open mesh design.

Another trap that utilizes the upward movement of moths is the wind-oriented or Brownsville trap, designed by Raulston, Sparks, and Lingren (1980) for the capture of live moths (Fig. 4.3i). It consists of a wind vane, a sheet steel, hardware mesh body, and a plywood-collecting container mounted on a pivot so that the open front faces downwind towards the pheromone plume. This trap is highly efficient (up to 42%), and is better than the Texas trap in wind speeds of over 9.9 km/h (Hendricks, Perez and Guerra, 1980). However, it is expensive and can be damaged by high winds. It has not been used in Australia to date.

III. Factors Affecting Trap Catches

1. Trap Placement and Spacing

The number of traps required and their placement in a given area depends on whether quantitative or qualitative results are needed. A single light or pheromone trap might suffice for a presence/absence study, but a network may be needed for quantitative work.

Moths congregate in areas where suitable hosts are present. Traps placed in such areas will give more reliable results than those placed outside them. In cotton, pheromone traps placed 40 m within a crop gave about double the catch of those placed on the edge (Wilson and Morton, 1989). Catches in low-wattage blacklight traps may be higher if placed within a crop, but the greater range of mercury vapor lights makes their positioning less critical. All trap types may suffer a bias if placed near a crop that is more attractive than the one being sampled. For example, a trap placed in cotton adjacent to silking maize will show a disproportionately high *H. armigera* catch. Other anomalies may occur where one crop is attractive for egg laying, but an adjoining crop is more attractive for feeding (Topper, 1978). This is unlikely to occur in large monocultures provided traps are located within the crop, but is a problem in small-scale farming areas in India (D.R. Dent, personal communication).

The condition of the crop is also important. Favorable conditions for egg laying may result in a higher level of oviposition than would be expected from a given moth population. Conversely, a substantial moth population may be observed in a crop with little oviposition. This often occurs in autumn cotton crops that have matured and become unattractive for laying. The height of the crop is important for pheromone traps. It is recommended that Texas traps should be adjusted continually so that the pheromone plume is released just above the canopy (Fig. 4.3h).

When pheromone and light traps are used together, or when traps are networked, the question of trap spacing arises. The effective radius of large light traps is of the order of 100 to 200 m and there may be some influence out to 1000 m (Hartstack et al., 1971). The range of small traps with UV tubes is presumably much less, whereas that of pheromone traps is highly dependent on wind speed. The response to pheromones is reduced markedly by even low levels of UV or visible light (Lam and Baumhover, 1982). These considerations suggest that if possible, all light and pheromone traps should be separated by not less than 100 m, or 200 m in the case of large light traps. For sites where a small light trap and pheromone traps for *H. armigera* and *H. punctigera* are combined, an appropriate layout is a triangle with sides of about 100 m.

2. Weather

Trap catches can be influenced greatly by wind speed, temperature, humidity, moonlight, and rainfall. These variables interact, and are usually auto-correlated. Moreover, the interactions may vary with population density. This makes it difficult to separate the influence of one weather variable from the others, and to correct the catch for their effects (Morton et al., 1981). Analyses based on hourly values may vary from those based on whole night values because inactivity in unfavorable periods may be compensated for by increased activity during favorable periods of the night. On the other hand, problems with hourly analyses arise because traps exploit behavioral responses that have underlying circadian rhythms. As temperature, wind-speed, and humidity depend on the time of night, it is hard to determine whether trap catches are related to the environment or to the behavioral pattern (Dent and Pawar, 1988).

Pheromone and light trap catches increase between 14°C and 26°C (Morton et al., 1981). Light trap catches cease below about 12°C, but there is no evidence of a lower temperature threshold for pheromone trap catches, and substantial catches have been recorded on nights with a minima of 5°C or less (Rothschild et al., 1982; A.G.L. Wilson, unpublished). The greater effectiveness of pheromone traps compared to light traps at low temperature is most noticeable in spring and autumn. The relation of spring pheromone catches to oviposition has not been researched, but in the USA, spring pheromone peaks predict subsequent generation peaks much better than do light traps (Hartstack et al., 1978).

The influence of humidity is not well understood, partly because it is often correlated with temperature. Dent and Pawar (1988) found that in India, peak light-trap catches of *H. armigera* were obtained when the humidity exceeded 75%, but noted that this might have been due to the time of night rather than humidity per se. Observations with pheromone traps indicate a highly variable response to humidity (Dent and Pawar, 1988; A.G.L. Wilson, unpublished).

Substantial but different responses to wind speed have been found for pheromone and light traps. For light traps, Morton et al., (1981) found an exponential decline in catch at wind speeds above 1.7 m/s. The effect was more pronounced for *H. punctigera* than for *H. armigera*. Dent and Pawar (1988) found similar patterns, although some moths were caught at wind-speeds of up to 4.7 m/s. Conversely, for pheromone traps, a minimum threshold of 0.7 m/s for moth catches has been demonstrated (A.G.L. Wilson, unpublished) and Dent and Pawar (1988) found that maximum pheromone trap catches occur when wind speeds range from 1.6 to 3 m/s. The difference between light and pheromone traps presumably is related to the requirement for some air movement for plume formation in the latter. Wind direction can influence catches also. With pheromone traps, the catches are greater on the downwind side of crops (Rothschild et al., 1982; A.G.L. Wilson, unpublished).

The influence of moonlight is likely to be complex. In many insects, there is a trimodal pattern of activity in the lunar cycle, with peaks at full moon and either side of new moon (Danthanarayana, 1986). Evidence for *H. armigera* and *H. punctigera* (Persson, 1976; Morton et al., 1981; Dent and Pawar, 1988) and *H. zea* (Nemec, 1971) indicates lower catches when the moon is bright. The latter author suggests that a lunar cycle of moth emergence exists, but it is difficult to verify this when the efficiency of light traps is lower on moonlit nights (Bowden, 1982; Dent and Pawar, 1988).

Morton et al. (1981) suggested that the effect of moonlight was greater for *H. armigera* than *H. punctigera*, a trend also apparent in the data of Persson

(1976). Observations during the 1985–1986 season when night skies were consistently clear showed a marked depression of catches in bright moonlight for both species in pheromone as well as light traps (Wilson and Bauer, 1986). In contrast to these observations, increased pheromone trap catch of *H. virescens* at full moon is reported in the USA, although oviposition declines at the same time (Hartstack et al., 1978). In India, moonlight had no marked effect on pheromone trap catches of *H. armigera* (Dent and Pawar, 1988).

There is little information regarding effects of rainfall, which in any case is likely to be correlated with temperature, humidity and cloud cover. However, D.B. Stuart and P.C. Gregg (unpublished) found that peak catches in pheromone traps in nonirrigated areas often occurred on nights when some rain fell.

3. Insecticide Application

Some spray applications have a major effect on the catch of moths in traps within or adjacent to the sprayed area. An average reduction in catch of about two-thirds was detected by Morton et al., (1981) following applications of DDT to cotton. Similar reductions in catch have been found in both light and pheromone traps following the application of synthetic pyrethroids (Wilson and Morton, 1989). Such reductions in catch may reflect correctly mortality of moths due to spraying and curtailment of oviposition within individual crops, but they may distort estimates of generation frequency and abundance within a broad area. The distortion may persist for up to 7 days because as well as knockdown and foliage contact effects, pyrethroids may be repellent to moths. Other insecticides may have less drastic effects on trap catches, but more work is required.

4. Competition Effects

Pheromone trap performance decreases as the numbers of female moths increase. This has been termed the *female competition effect*. It has been shown in mark-recapture studies that feral females compete with synthetic pheromones thereby reducing trap catches (Hartstack and Witz, 1981). The effect may explain why pheromone trap catches are higher than are light trap catches early in the season, whereas later the reverse may occur (Hendricks et al., 1973a). Another problem may be that trap catch is influenced by the proportion of calling females. Ovipositing females may not be interested in mating, so that an imbalance between males and receptive females occurs, leading to a pheromone trap peak after oviposition rather than before it (Sparks et al., 1982). In Australia, the best correlation appears to be between oviposition and trap catches just before or at the same time as oviposition (Rothschild et al., 1982; A.G.L. Wilson, unpublished), but this correlation is often poor.

IV. Trapping Procedure

1. Recommended Standard Designs

Each trap has advantages and disadvantages, so it is impossible to select the single best design for every application. The growing plethora of designs means that unless some attempt is made at standardization our data base will become ever more fragmentary.

In the case of light traps, there is clearly a need for two standard designs, one large and one small. For the former, we suggest the Robinson trap equipped with a Philips HPL-N-125W bulb, as used by the Narrabri Agricultural Research Station (Fig. 4.2a). For the latter, Pennsylvania traps fitted with 6-W BL tubes are probably best (Fig. 4.2d).

For pheromone traps, the 75:50 Texas trap made of hardware cloth or sunlight-resistant shadecloth represents the best compromise at present between ease of use and catch efficiency for both species. The dry funnel trap is in widespread use, but this should be continued only where ease of use is a critical factor and only qualitative data are required.[3]

2. Tips for Trap Operation

The amount of information gained from trapping moths will be in direct proportion to the effort put into collection and processing of catches. Some considerations for trap operation are discussed below.

A. Light Traps

To check on the operation of the trap, there is no substitute for regular inspections at night. Do not assume just because moths are caught that the trap is functioning normally. Fluorescent tubes become duller over time, and batteries may lose their charge. Also check that the timing mechanism is operating over the correct time span.

A little effort made in preserving the catch in good condition will save much tedious dissection. The use of appropriate containers and/or materials, such as egg cartons or crumpled newspaper in the containers is essential. If specimens are collected dry, the collection interval should be no more than 24 h. For wet preserved material, ensure that the alcohol has not been weakened by evaporation or rain.

If there is any doubt about identification the specimens will have to be dissected and identified on the basis of differences in the genitalia (Common, 1953). A conservative approach is best, as abrasion marks often can look remarkably like the pale wing marking of H. armigera.

---

[3]Drawings and designs for each of these standard traps can be obtained from P.C. Gregg.

B. Pheromone Traps

Maintain lures in the deep freeze until required, and change them regularly once per month.

Moths are highly sensitive to minute amounts of pheromone and contamination of one species lure by another may result in loss of specificity. Lures should be stored separately and handled with disposable gloves or forceps. New gloves should be worn for each species, or separate marked forceps should be used.

Traps should be checked regularly, daily if possible, but otherwise at a maximum of 3-day intervals.

Chapter 5

Estimating the Abundance of Adults and Immatures

I.J. Titmarsh, M.P. Zalucki, P.M. Room, M.L. Evans, P.C. Gregg, and D.A.H. Murray

I. Why Estimate Abundance?

At the single field or farm level, pest managers who consider how *Heliothis* spp. affect their crops need current information on *Heliothis* numbers on which to base management decisions and assess the efficacy of past control tactics (Wilson and Room, 1982). At the farming district or regional level, population researchers who consider how crops and control tactics affect *Heliothis* must assess population densities of *Heliothis* developmental stages continually. From these data, calculation of current population numbers may be used to estimate, for example, birth and survival rates or assess the worth of alternative control strategies.

In all but the smallest areas of interest (for example, a leaf) we can never know exactly how many *Heliothis* are present, and we must estimate their density by sampling. Such estimates may be either absolute, expressing *Heliothis* densities as numbers per unit area; or relative, measuring population in "unknown" units that allow only temporal and/or spatial comparisons. The latter are either catch per unit effort or trapping methods, the results of which can depend on factors other than population size (Southwood, 1978).

The distinction between absolute and relative methods is not always clear, as absolute estimates are usually less than 100% correct and relative estimates can be very accurate or adjusted to give absolute estimates (Southwood, 1978).

A variety of sampling techniques is available, and these must be compared before choosing the one appropriate for a given situation. For instance, no scout will be able to afford to use more than one technique when sampling for decision making. Generally, it is accepted that no one technique will best sample all stages, species, and relevant plant parts in a crop (Mayse, Kogan, and Price, 1978a; Marston, Dickerson, Ponder, and Booth, 1979; Bechinski and Pedigo, 1982). Clearly some compromise must be made where practical considerations are of prime importance, so that the final decision will be determined by the type of information required and the ability to carry out the sampling realistically.

II. How Many Samples?

1. Number of Samples

The accuracy of population estimates will improve as the number of samples increases. Unfortunately, this will also take more time! As resources available to researchers are usually limited, an amount of inaccuracy must be accepted within the population estimate — usually a balance between the cost of sampling time and effort and the "cost" of making a wrong decision.

The number of samples needed depends on the precision required, the pattern of insect dispersion, the expected value of the mean and the variance in the data. The negative binomial distribution has been used frequently to describe insect dispersion, at least within relatively homogeneous crops and when such different variables as life-stages, species, and plant parts have to be sampled simultaneously (Wilson, Room, and Bourne, 1983). The variance depends on the habitat/crop being sampled, the phenology of the plant and the insect, the sample method, and the history of the area being sampled. An accurate assessment of the number of samples needed at a particular time and place can only be obtained by taking a set of preliminary samples. Usually, this is prohibitively expensive and assessments are made from data gathered in previous years with the same sampling method on the same host-plant at the same time of year, ideally in the same geographic location. In most projects, different variables are sampled simultaneously and it must be decided which are the most important and will be used as the basis for calculating the number of samples needed.

A general equation for n, the number of samples needed to give an estimate of (μ) the mean, whose $(1 - \alpha)$ half-confidence interval is a given proportion D of the mean, was given by Karandinos (1976) as

$$n = (Z_{\alpha/2}/D)^2 \cdot \sigma^2/\mu^2 \tag{1}$$

where $(Z_{\alpha/2})$ = the upper $\alpha/2$ point of the standard normal distribution and σ^2 = the variance.

In practice, estimates of σ^2 (S^2) and μ ($\bar{x}$) are used and Student's t is substituted for Z because the mean and variance are unknown. The negative binomial is usually a better mathematical model of insect dispersion than the normal distribution (due to the tendency of insects to be clumped rather than randomly distributed), and square root transformation of the data is needed to make the estimate valid (Wilson and Room, 1982).

High levels of precision are expensive because n increases as a function of D squared. For example, n must be multiplied by 100 to decrease D from 1.0 to 0.1, and by 100 again to decrease D from 0.1 to 0.01. Similarly, low population densities require more samples than do high population densities. In counting eggs of *Heliothis* by eye in cotton, to obtain population estimates with 90% probability of being within 10% of the true mean, 17,000 plants must be examined when there are 0.1 eggs per square meter compared with 700 when there are 5.0 eggs per square meter (Wilson and Room, 1982, and below).

There are techniques, such as sequential sampling (Sterling, 1976), that allow sampling to cease when a given level of precision has been reached. Sequential sampling is being used in pest management programs to halt sampling when the mean is significantly above or below a single, predetermined level. In research, we often need to estimate means of all magnitudes with the same relative precision, and the use of sequential techniques would require large numbers of tables to be carried. Alternatively, calculations could be updated with each additional data point by field- portable computers.

Binomial sampling (presence/absence) has been used to reduce sampling effort in pest management. Population densities of insects have been cross-calibrated with proportions of infested plants (Wilson and Room, 1983) so that, in future sampling, numbers of individuals do not have to be counted. The major drawback is that the reliability of population estimates decreases rapidly above 50% of plants infested, until the estimates are almost meaningless when all plants are infested.

2. Dispersion of Samples

To obtain unbiased estimates, samples should be collected at random. Tables of random numbers can be used to select, for example, the crop row and distance down it at which the sample will be collected. This can be a time-consuming procedure and assumes that the universe (field, plot, etc) being sampled is uniform.

In practice, in such visually homogeneous areas as large crop fields it is more usual to use procedures for selecting samples that are systematic at the large scale and have a degree of randomness at the finer scale. This ensures that no areas remain unsampled and minimizes the time spent moving between sample points. In much of the sampling of immature *Heliothis* spp. in cotton, for example, rows were selected at regular intervals throughout the field, but starting at a different position on each occasion, and individual plants were selected after walking a fixed number of paces before counting off 7 stems. The lengths of paces varied between individual samplers and their moods, introducing a degree of randomness, and counting a set number of stems precluded biased selection of plants based on size or other characteristics.

3. Frequency of Sampling

How frequently to sample is dictated by the use to which the results are put. If it is desired to sample a particular *Heliothis* developmental stage, then clearly the sampling interval needs to be less than the duration of the stage and much less if several samples are to be taken during the transit of the particular stage.

In the absence of immigration, if only one sample were to be taken per generation, then the sampling interval need not be less than 42 days, as this is the fastest time in which a generation can develop in most seasons [at least for the Namoi Valley in NSW, Australia, according to Room (1983)], it is essential to sample at least every 3 days in summer if any estimate of egg input is to be made.

A feel for how often to sample is gained with experience. An indication of the durations of the egg and pupal stages, of development from oviposition to 8-mm larva, and of oviposition peak to oviposition peak is given in Room (1983).

4. Night Sampling

Even though *Heliothis* moths and various predators of *Heliothis* adults and immatures are only active at night, it is only in the last decade that any attempt has been made to study nocturnal behaviour (see Chapter 11). Little of this work has been conducted in Australia.

Tools for nocturnal sampling include head lamps, nightvision devices (usually expensive and of military origin), pheromone/light/malaise traps, mating tables, actographs, and radar (Lingren, Sparks, and Raulston, 1982; see Chapter 11).

An apparent problem with white light is moth avoidance or attraction, however, this is largely dogmatic. Newly emerged adults and mating pairs cannot escape from light and adults engaged in other behavior such as feeding generally continue the activity even in the presence of light (Lingren et al., 1982). Once having located an insect, any behavioral disruption may be averted by covering the beam with a red filter or redirecting the beam so that the insect appears in the fringe of illumination (Lingren et al., 1982).

The nocturnal behavior of larvae is not known, but cooler night temperatures mean less physiological development and possibly reduced mobility.

III. Estimating the Abundance of Adults

1. Absolute Methods

A. *Mark-Recapture*

If a known number of marked individuals is released and allowed to disperse in a population that is then sampled, the unknown population size (N) will bear the same relationship to the sample size (s) as the number of marked individuals released (M) does to the number recaptured (r), that is,

$$\frac{s}{N} = \frac{r}{M}$$

The simplest application is the Lincoln Index, where a single release and a single recapture are employed, but more complex methods involving multiple releases and samples are available. Detailed descriptions of all these methods and the assumptions underlying them, with formulae for calculating population size and variance, are provided by Seber (1973), Southwood (1978), and Begon (1979).

Heliothis adults may be marked by a number of methods including spraying with the dye rhodamine B (Hartstack et al., 1971; Haile et al., 1975), through feeding Calco Red N-1700 and Oil Soluble Blue II dyes in larval diets (Hendricks and Graham, 1970), and labelling with either radioactive ^{32}P in larval (Snow et al., 1969) or adult food (Room, 1977) or with rubidium through the application of rubidium chloride to host plants of the larvae (Graham and Wolfenbarger, 1977; Graham et al., 1978a,b).

All mark-recapture methods involve a number of assumptions, some of which are quite dubious when applied to adult *Heliothis* spp. The most important are:

1. that moths are not affected by being marked;
2. that marked moths are mixed completely and distributed randomly within the population at the time of sampling so that the probability of recapturing a marked moth is the same as capturing an unmarked one;
3. that the population is marked and sampled randomly with respect to age and sex. Failing this, any biases in the released insects should be reflected in the sample(s) and appropriate allowances made in calculating the total population size;
4. that all moths, regardless of their position in the habitat, are equally available for capture.

Limitations to these assumptions have been recognized in experiments (for example, Hendricks, Graham, and Raulston, 1973b) intended to yield crude estimates of *H. zea* and *H. virescens* dispersal and migration. To quantify population size will require careful compliance with the assumptions of the method.

Considerable attention must be given to the possibility that assumption (1) may not be satisfied with marked *Heliothis* moths. Where laboratory reared moths are used, they may differ in viability or behavior from wild moths (Hendricks et al., 1973b; Haile et al., 1975). The results of Ramaswany, Roush, and Kitten (1985) indicated no such problem when *H. virescens* and its backcross hybrids were used, though it seems likely that irradiated *H. virescens* were less competitive (Hendricks et al., 1973b). It may be possible to overcome such problems by trapping and releasing wild moths, perhaps by means of a self-marking trap as developed for stable flies by Hogsette (1983). The reaction of *Heliothis* to being trapped is unknown and it is possible that even a brief and nontraumatic capture can alter the probability of recapture profoundly, as demonstrated in a tropical Papilionid butterfly by Singer and Wedlake (1981). There is a need for work on these aspects with *Heliothis*.

Assumption (2) poses problems because there is little information on the rate of dispersal of marked moths into the general population. Proshold, Raulston, Martin, and Laster (1983) suggest that there is an incomplete mixing of backcross insects in the premating period, if they are released as adults.

Assumption (3) involves some obvious difficulties, because the most convenient methods for capturing moths to be marked and subsequently sampled are light and pheromone traps. These have pronounced biases with regard to age, sex, and mated status (see Chapter 4). However, the biases are probably sufficiently understood to enable some corrections to be made. Hartstack et al. (1971) used mark-recapture methods to study the effective range of light traps. They present a method whereby their results could be used in conjunction with Lincoln Index procedures to estimate population density.

There are major difficulties with (4) because of the wide host range of *Heliothis*, the capacity of moths to disperse at least several kilometers within 24 h of marking (Hendricks et al., 1973b), and their ability to aggregate in attractive areas (Snow et al., 1969). Therefore, an extensive sampling programme is required if the population of a substantial area is to be measured. Snow et al. (1969) used 250 light traps to estimate the population of *H. zea* on St. Croix, an island of only 218 km^2, but were still unable to obtain an accurate figure.

An interesting variant of the mark-recapture method is that of Kelker (1940, in Southwood, 1978), which relies on the selective removal of individuals with natural marks. Gender is one such mark, and distortions in the sex ratio caused by removal of known numbers of one sex can be used to estimate population size. This raises the possibility of measuring *Heliothis* adult populations in a small area by appropriate operation of highly effective pheromone traps (inverted cone types — see Chapter 4) and light traps in the same locations. Difficulties with (3) and (4) would arise obviously, but it might be possible to overcome them by simultaneous operation of those separate light and pheromone traps in areas nearby, but out of trapping range.

Although mark-recapture experiments are costly in terms of the limited data produced per unit effort, the results obtained provide information additional to population size estimates, in for example, the distance and direction between mark and recapture sites and the change in condition between mark and recapture. Obviously the greater the number of recaptures, the more useful this information will be.

2. Relative Methods

A. *Trapping (with Correction)*

Light traps and, more recently, pheromone traps have been the most widely used traps in *Heliothis* studies (see Chapter 4). Other trap types such as the emergence cage, malaise trap, and tow net can be of special interest in particular circumstances (see Chapters 9 and 10).

As the size of the nightly catch is determined partly by the number of available moths, partly by the effect of such environmental factors as temperature and wind speed on moth activity, and partly by nocturnal changes in insect behavior (Morton et al., 1981), it is hardly surprising that trap efficiency varies between *Heliothis* species, between nights, and between sites.

Mark-recapture experiments have been used to quantify the efficiency of different traps (Hartstack et al., 1971). Laboratory-reared or wild moths marked with rhodamine B dye on the wings, pronotum, or ventral abdomen were released at various distances and directions from trap sites. The proportion of marked animals captured provided an estimate of trap efficiency. Recapture percentage ranged from ca. 40% for moths released 10 ft from a light trap to 2% for releases 270 ft away. Hartstack et al. (1971) provide equations of trap performance for estimating the percentage of adult *H. zea* that light traps will remove. Equivalent work on Australian *Heliothis* is lacking.

In most instances, only one light trap has been operated per field compared with several pheromone traps. This means the sample variance of light traps cannot be calculated. Where practicable, it will always be advantageous to run a grid of traps.

B. *Walk and Count or Line Transect*

Walk and count or transect methods for estimating adult *Heliothis* have not been used widely (see Greenstone 1979, on the application of catch per unit effort in entomological studies).

Here the observer walks a line transect at a constant speed through a habitat (for example, a cotton or soybean field, etc) and records the number of moths seen. The number recorded will depend on the actual population density, speed of movement of the moths, the observer's speed, and the distance over which the moths can be seen — this will depend on the ability of the observer and on the habitat. A static or dynamic model is used then to make density estimates, depending on the time of day the transect count was taken.

Generally, *Heliothis* adults are inactive during the day. By walking through a field during the day the observer will disturb and flush out some proportion of moths from some area of the habitat. If, as Southwood (1978) points out, this proportion is constant, then the number of moths seen gives an index of the absolute population. If this proportion and the area covered are known, then the absolute population density can be estimated. If the efficiency of flushing varies, then only a relative measure of density is obtained.

Heliothis adults are active during the night, although the level of activity varies with time of night, ambient conditions, and the age and sex of the moths (Persson, 1976). A line transect taken at night will require a "dynamic model" to obtain a population estimate. The basic formula is

$$D = \frac{N}{2rV}$$

where D is density per unit area; N, the number of moths seen per unit time; r, the radial distance over which moths are observed, and V, the average velocity of the moth relative to the observer, that is,

$$V^2 = u^2 + w^2$$

where u is the average velocity of the observer and w the average velocity of the animal. Of course, the units for velocity, distance, and time need to be consistent (Southwood, 1978).

There are a number of difficulties in using this method, one of which is measuring the speed of active and flushed moths. Due to the variation in activity within a night, transects will need to be taken at the same time if comparisons between nights are to be made. Furthermore, activity will vary markedly between nights due to weather conditions. Numbers counted may reflect temperature, wind, and moonlight conditions as much as actual population size. Night transects will also require some way of seeing moths, however, any method that uses additional lighting may affect moth activity (above, Chapter 11). A check on species identification will be required as many similar-sized moths may occur in the one area.

C. Bait Sprays

A possible alternative to line transects is to use a bait spray applied to some area of habitat, such as along a number of rows of a crop. A bait such as honey or sucrose (both feeding stimulants), cellulose acetate (a sticking agent), and a killing agent (for example, methomyl insecticide) has been tried with some success (M.J. Rice and M.P. Zalucki, unpublished). Such a bait kills male and female moths that come across the bait and feed. These moths can be collected the next morning on the ground. The number of moths killed will depend on the area baited, the population density, and the proportion of moths feeding. The number collected will depend on how efficiently dead moths can be found. This will depend in part on how far baited moths fly before dying, and what fraction are removed by ants, predatory beetles, and so forth. As a population estimation technique for *Heliothis*, "bait-lining" requires further research.

D. Insecticidal Knockdown

Insecticides provide a tool for measuring moth populations although the cost of treating and counting from large areas of habitat has deterred use of this method.

Adulticidal sprays drop moths onto the ground, or onto ground sheets positioned earlier, that can be retrieved or on which moths can be counted. The numbers of moths killed and collected will depend again largely on the limitations discussed under bait sprays (above), but it is essential to choose an effective and nonirritant adulticide (such as profenofos) to overcome problems of repellency and insecticide-resistance. Otherwise, moth numbers may be underestimated seriously.

E. Radar

Several types of radar are available, their greatest strength being that the very large volume scanned (about 3×10^7 m³) ensures that measurements are possible even when insect densities are low (V.A. Drake, personal communication).

An important limitation of the radar technique is its inability to provide specific identifications of detected insects. Usually, target identity must be established by more conventional methods such as trapping, the most satisfactory being direct sampling using tow nets (Farrow and Dowse 1984, and Chapter 10) carried out simultaneously with the radar observations.

IV. Problems in Estimating Abundance of Adults

Many of the difficulties in estimating adult abundance by either absolute or relative techniques are related to the vagility of adults and their large population sizes in agricultural areas. Adults are active mainly at night and they can move large distances both within and between nights. Activity will vary with the age of moths, relative density of the alternate sex, availability of adult food, wind speed and direction, temperature, cloud cover and moon phase among other things. All of these factors will affect the efficiency of trap capture and sampling.

The major difficulty with the use of mark-recapture methods for *Heliothis* adults is the very low recapture rate. This may reflect the high vagility of moths and hence, the transient nature of a population marked within an area; or the population may be so large that the fraction marked and sampled is very small. These two possibilities are related to the area over which samples are taken. If this area is small relative to the mobility of animals, then one is measuring effectively a subpopulation of a much larger population. Unless very large numbers of moths can be taken and marked effectively, then mark-recapture methods are of limited use in population estimation. Even if such sampling were possible, the other assumptions of the mark-recapture method (above, and Begon, 1979) need to be taken into account. For instance, the high vagility of moths may result in the violation of the assumption that once a moth has left an area it does not return.

The difficulty with using relative methods to estimate abundance is that the estimate varies with changes in many variables — age of moth, time of night, activity of other moths, and environmental conditions. If the effect of these variables can be measured then the abundance estimates can be corrected for their effect. Such correction does not necessarily give us an absolute estimate because the traps may be selective. Simply capturing animals under a variety of conditions and correlating the catch with ambient conditions does not answer the problem. We need to know the efficiency with which a trap samples an area and the responsiveness of moths to trap stimuli, particularly if this response to attractants differs between moths of different ages, sexes, and genotypes. If this information is not obtained, then the interpretation of trap catches must remain limited to presence/absence and some indication of the timing of population peaks.

One way of overcoming the problem of high vagility is to utilize a network of traps over an extensive area. These traps need to be cleared and catches monitored frequently — preferably on at least a daily basis. Photoreceptors and portable, cheap data loggers coupled with specific attractant traps (for example, male pheromone traps) make such a simultaneous trapping effort feasible. The pattern of catches over such a trapping grid can indicate the direction and intensity of migration as opposed to local emergence (Campion, Bettany, McGinnigle, and Taylor, 1977; Pedgley, 1986). To do this will also require monitoring of the local microclimate as well as synoptic weather patterns (see Chapter 10).

Unfortunately, the only attractants available that are specific for *Heliothis* are the male attractant pheromones. These sample males only and the pattern of catch is often confounded by the competing effects of virgin females (see Chapter 4). Although light traps could be used for such a monitoring grid, the effort and cost required to clear and sort the contents of a light trap on a daily basis would be prohibitive. There is an urgent need for the development of a specific female lure.

V. Estimating the Abundance of Eggs and Larvae

This will usually involve a single technique for fields of any particular crop. At the whole farm and the regional level a number of sampling techniques may be required to cope with the diversity of host plants used by *Heliothis*.

1. Absolute Methods

A. Mark-Recapture

The limited ability of eggs and larvae to disperse seemingly precludes any application of this method for absolute estimates of population. However, the recapture of individuals through time can measure the natural mortality of a marked cohort as the decline in successive absolute estimates of the cohort's numbers (see Chapter 6).

A method for producing radiolabelled (marked) *Heliothis* eggs and larvae for such studies has been developed by Room (1977), although any application of the method to date has been trivial in the context of measuring abundance.

B. Fumigation Cage

In Pedigo, Lentz, Stone, and Cox (1972), a wooden receptacle was clamped together at the base of plants 1 day prior to sampling. Each census of approximately 0.3 m was made by cautiously approaching the sampling site and quickly setting a polyethylene cage with removable lid over the plants and into the receptacle. Petri dishes of calcium cyanide were introduced to the cage, left for about 10 min and then removed. Subsequently, the cage lid was removed, plants were shaken vigorously, and then, with the cage removed, the dead insects were recovered from the receptacle. Catches were stored and later sorted in the laboratory.

C. Cylinder

As for the fumigation cage, a volume of plant material is enclosed within a covered cylinder and an insecticide is introduced to provide knockdown of insects that are then collected. Kretzschmar (1948) placed base plates of transparent plastic stretched on wooden frames beneath plants the day before sampling. Cylinders constructed of plastic stretched over a wire frame were placed over the base plates at sampling time and potassium cyanide was introduced. The base plates ensured easy collection of the catch from soil level.

D. Clam Trap

This trap is modelled on bivalve principles where trap halves, hinged on one side, close over portions of a plant row from each side of the row. Plant stems are severed either by the closing of the "jaws" or by hand once the trap is closed. The killing agent is introduced and the plants and/or insect catches are containerized for subsequent scrutiny in the laboratory. An example is Mayse, Price, and Kogan (1978b).

E. Whole Plant Removal

Whole plants are containerized and removed from the field for later inspection in the laboratory. Bechinski and Pedigo (1982) carefully placed plastic bags over 0.3m lengths of row, cut the plant stems at the soil surface, introduced ethyl acetate as a killing agent, and then froze the samples for later processing. Insects were recovered by rinsing samples in soapy water and filtering the wash. Wilson and Room (1982) severed the mainstem of a single plant at ground level first, then the plant was placed in a plastic bag and ethyl acetate was added. On return to the laboratory, plants were dissected and insects counted (see Chapter 7).

2. Relative Methods

A. Direct Observation

Here examination may be of the whole plant or, where appropriate, of particular structures only (for example, fruit). The plants are traversed visually in a systematic way, branch by branch, and from the top down or vice versa. The number of insects per unit area or per length of row at each sampling date is determined by recording the numbers of plants per unit area or per length of row.

B. Sweep Netting

The diameter of the sweep net is usually specified. The bag should be white to assist in the detection of insects in captured material, and be of heavy duty, but fairly fine material. Usually, insect numbers are referred to as per "sweep," but can be quantified per unit area if each sweep, for example, corresponds with and is the same length as a 1-m pace. The type of sweep used and insect behavior can affect the percentage capture of insects in certain categories. For instance, if large larvae tend to drop off the plant when disturbed, then an upward sweeping motion may be most suitable. Similarly, any basking by larvae at the top of the plant in early morning will result in higher sweep net catches at that time (Mabbett, Dareepat, and Nachapong, 1980).

C. Plant Shake or Ground Cloth

This involves holding a predetermined length of row and shaking it in a set manner into a collecting receptacle. The length of row, manner of shake, and receptacle may be varied according to situations and operators, however, a plastic sheet on the ground facilitates the sorting and removal of insects by preventing them from getting a good grip, is easily cleaned, and any excessive length may be thrown over the adjacent row to prevent the significant number of insect escapes that can occur through this region (Marston et al., 1979). For similar reasons the ground cloth is usually wider than the length of row being sampled.

D. Insecticidal Knockdown

This method is similar to the ground cloth except that once the ground sheets are in place an insecticide providing quick knockdown of larvae is applied, some time after which plant shaking and insect collection proceed. Clearly this method is unsuitable for estimating egg numbers.

3. Comparisons of Sampling Techniques

In Australia, researchers have compared sampling techniques for catch efficiency, reliability, and cost in cotton (Wilson and Room, 1982) and soybean (Evans, 1988).

In cotton, Wilson and Room (1982) assessed 3 seasons' population densities of 26 arthropod species or groups in six fields by

1. direct observation of single plants placed in plastic bags in the field and processed in the laboratory (an absolute technique);
2. direct observation of individual cotton plants in the field; and
3. laboratory inspection of sweep net catches from 30 m of crop row.

Sweep net sampling was least expensive at any level of reliability, but consistently gave the lowest population density estimates (Table 5.1). In addition, nets could not be used to detect eggs and early instar *Heliothis* or to sample cotton fruit. In comparison, although bagging samples was the most expensive method at any reliability level, fewer bags than visual (direct observation) samples were required for a given level of reliability in the population estimate.

Table 5.1. Population Densities of *Heliothis* in Cotton Fields

	Mean Population/m^2			
	Bag	Visual on "Bag Days"	Visual on "Sweep Days"	Sweep
	No. of Plants			
Stage	1,728	3,368	7,944	318,500
White eggs	1.113	0.974	0.282	0
Brown eggs	0.881	0.984	0.373	0
Very small larvae (<3 mm)	1.150	0.260	0.147	0.001
Small larvae (3–7 mm)	0.681	0.433	0.175	0.005
Medium larvae (7–19 mm)	0.142	0.139	0.084	0.005
Large larvae (>19 mm)	0.062	0.063	0.051	0.003

Means for three seasons estimated by three sampling methods used in the same fields on the same days (Wilson and Room, 1982).

An all season average time, including walking time between sample sites and later laboratory sorting of samples, was 2 min for a visual sample, 7 min for a bag sample, and 8 min for a sweep sample. Optimum numbers of samples for cotton arthropods and costs in person-hours of obtaining them are reproduced in Table 5.2.

In soybean, Evans (1988) sampled arthropods using

1. fumigation cages (absolute technique);
2. direct observation of individual plants in the field;
3. field inspection of sweep net catches; and
4. field inspection of ground cloth collections from 1.5 m of crop row.

Median insect numbers estimated by direct observation were usually closest to fumigation cage estimates (Table 5.3). Numbers collected by ground cloth were next in magnitude and much larger than those detected by the sweep net, which again gave the lowest estimates consistently. Although the ground cloth collected a smaller percentage of larvae as population size increased, the relationship was reliable and relative numbers were correlated significantly with absolute numbers, and hence, conversion to absolute estimates was possible.

The fumigation cage was the least, and ground cloth and sweep net were the most cost-efficient of the techniques. Cost-efficiency was assessed as the cost expended to produce an estimate of a specified accuracy, and so is unrelated to the number of samples taken. Median relative cost-efficiencies

Table 5.2. Optimum Numbers of Samples for *Heliothis* in Cotton and Costs in Person-Hours of Obtaining Them[a]

Sample Method and Stage	Optimum Numbers of Samples				Cost in Person-Hours			
	Population Density							
	0.1[b]	1[b]	2[b]	5[b]	0.1	1	2	5
Bag								
Total eggs	17,000	2,600	1,500	700	2,000	300	170	80
Visual								
Total eggs	29,000	3,700	2,000	890	1,000	120	70	30
Small larvae	107,000	5,700	2,400	700	3,600	200	80	20
Sweep								
Small larvae	40,000	3,100	1,400	500	5,300	400	190	70

[a]$\alpha = 0.2$ and $D = 0.1$ (Wilson and Room, 1982). To find number of samples or cost for $D = 1.0$, divide by 100, see Eq. (1).
[b]Population densities per m2 as measured by the visual method; population density per plant = density per m2/15 plants per m2.

(Table 5.4) of the various techniques varied with year, site, and across the season. Midseason efficiencies were generally highest, reflecting highest median insect numbers.

Time taken to complete one sample was longest for the fumigation cage, shorter for the sweep net and shortest for direct observation or ground cloth. Sampling time also varied across the season, the longest times per

Table 5.3. Median *Heliothis* Larvae Numbers per Row-Meter[a] Collected During Three Crop Growth Periods[b,c]

Sampling Technique	Site/Year								
	Coominya 1979–1980			Kalbar 1979–1980			Kalbar 1980–1981		
	I	II	III	I	II	III	I	II	III
DO	2.36	0.60	0.33	0.36	0.71	0.35	0.60	1.51	0.48
GC	0.27	0.45	0.00	0.41	1.25	0.25	0.71	0.51	0.16
SN	–	0.00	0.01	0.03	0.03	0.01	0.01	0.06	0.01
FC	0.83	4.01	–	2.08	0.25	0.83	–	–	–

DO, direct observation; GC, ground cloth; SN, sweep net; FC, fumigation cage.
[a]Medians of weekly means (calculated on untransformed numbers of insects per sample) over the crop growth period specified.
[b]I, planting to preflowering; II, flowering to full seed; III, full seed to harvest; –, no sampling done in these periods.
[c]Evans, 1988.

Table 5.4. Median Relative Cost-Efficiency[a] * 10 for Sampling Techniques for Three Crop Growth Periods[b,c]

Sampling Technique	Site/Year				
	Kalbar 1979–1980		Kalbar 1980–1981		
	II	III	I	II	III
DO	0.64	0.86	1.71	1.33	1.61
GC	6.81	3.57	5.24	1.91	3.31
SN	2.43	1.88	1.98	10.76	2.23
FC	5.07	0.91	–	–	–

RCE, relative cost-efficiency; CV, coefficient of variation; other abbreviations as in Table 5.3.
[a]RCE = $1/[CV^2 * \text{cost (time in h)}]$, where CV was computed using means of untransformed data; the higher the RCE values, the more cost-efficient the technique; the figures presented here are medians of the weekly figures for the crop growth period specified.
[b]I, planting to preflowering; II, flowering to full seed; III, full seed to harvest; –, no sampling done in these periods.
[c]Evans, 1988.

sample being recorded midseason. The relative strengths of different sampling methods are compared in Table 5.5.

4. Problems with Different Host Plants

A. Sampling Crop Hosts

The close spacing of crop plants effectively means that the time spent searching for another plant is negligible. Any problem in sampling crop hosts usually occurs because of particular plant characteristics.

The first problem arises due to differences in plant stature. Clearly, a sampling program based on sweep netting may work well in such crops as lucerne, but may be completely inappropriate for sampling from sunflower heads. Similarly, soybean or cotton plants may be bagged and removed from the field more easily than 2- to 3-m high maize plants. Consequently, crop host will influence the choice of sampling method.

Second, during later growth phases plants such as chickpeas, cotton, and soybean can intertwine to such an extent that disentanglement cannot be attained without some plant damage. For sampling to continue will require either a change of sampling methods, or the establishment of fixed sampling sites by removing plants at either end of the row length that is sampled. Ideally, artificially created breaks should be similar in size to 'natural' breaks observed in other rows (Evans, 1988).

Table 5.5. Relative Advantages and Disadvantages of Techniques Used to Sample *Heliothis* Immatures

Criterion	Advantage[a]	Disadvantage[a]
Samples the whole plant	2,3,4,5,6,8,9	7
High degree of sampling precision	1,2,3,4,5,6	7
Calibration of relative sampling techniques	2,3,4,5	
Labor intensive and/or time-consuming	7	2,3,4,5,6
Preparatory work prior to sampling		1,2,3,9
Possible plant death/injury caused		2,3,7,8
Ineffective on small plants		2,3,4,7,8
Less effective on intertwined plants	5,7,9	2,3,4,6,8
May alter insect/plant population being sampled		2,3,4,5,9
Examination of residual vegetation required		2,3,4
Reveals spatial orientation of individuals	6	
Permits behavioral/interactional observations	6	
Minimum disruption to system under study	1,6,8	2,3,4,5,7,9
Operable in most weather conditions	1,5,6,8	2,3,4,7,9
Includes many plants in each sample	7	2,3,6,9
Suitable for binomial sampling	7,8	
Sample affected by plant age/insect behavior		7
Sensitive to operator's ability		6

[a]Absolute techiques: 1, mark/recapture; 2, fumigation cage; 3, cylinder; 4, clam trap; 5, whole plant removal. Relative techniques: 6, direct observation; 7, sweep netting; 8, plant shake or ground cloth; 9, insecticidal knockdown.

Another problem is the comfort of the sampler. Chickpeas are covered with malic acid and certain tobacco varieties exude irritant gums. Both can effect the human skin. Sorghum pollen will produce intense allergic reactions in many field workers. Sampling safflower via direct observation rarely inspires enthusiasm. Certain crops, or crops at certain stages of development, require that sampling methods be modified to improve worker comfort. This might include the provision of protective clothing in safflower, respirators or full protective headgear in sorghum, or devising a method for plant manipulation without human contact. For example, shaking chickpea plants onto a ground cloth can be achieved by holding a stick of the desired row meter length and using it in a horizontal position to beat the plants. Sampling is often unpleasant enough without having to suffer crop-induced malaise.

B. Sampling Noncrop Hosts

Sampling from weeds, pasture plants, and other noncrop hosts may present special difficulties because they are not evenly spaced as are crops. Both the hosts within an area and the *Heliothis* on these hosts may show a clumped distribution. Consequently, the use of an appropriate contagious distribution such as the negative binomial to describe animal dispersion is essential (Southwood, 1978).

If sampling is intended to compare the suitability of various hosts, rather than to estimate the absolute population density, timed sampling methods may be used. M. Esmail and P.C. Gregg (unpublished) used this approach to sample *Heliothis* on pasture plants and weeds. When a sufficiently homogeneous patch (0.5 m × 0.5 m) was located, the time required to search it visually was recorded. These times ranged from 8 to 25 min depending on the size and vegetative characteristics of the plant. Later samples on scattered plants were made by searching for an equivalent time. This method is valid as long as the time required to locate new hosts is negligible compared to the time required to search them.

Where the hosts are scattered, it may be useful to estimate their density as well as that of the insect in order to determine whether changes in the latter reflect real population changes or changes in the distribution and abundance of the host. Methods for estimating the density of plants are discussed by Goldsmith, Harrison, and Morton (1986).

VI. Estimating Abundance of Pupae

Pupae of *Heliothis* spp. are found at 1- to 10-cm depth in the soil. The prepupa constructs a characteristic emergence tunnel from just below the soil surface down to the pupal cell, an enlarged chamber in which pupation takes place. After moth emergence, the pupal tunnel and cell remain intact, and the pupal remnants can be used for species and sex determination.

Quadrats are suitable for use in pupal sampling. In row crops, quadrats 1 row-meter in length and the inter-row distance in width should be used. Each quadrat is placed along the plant row to cover either side of the plant row to the midpoint of the inter-row spaces. Wilson (1983) found over 70% of emergence tunnels located in irrigated maize crop residues were within 5 cm of the center of the ridge and less than 3% were further than 20 cm from the center. In rain grown crops where ridging is less prominent, pupae are found further from the plant row (D.A.H. Murray, unpublished). Meter square quadrats are suitable for sampling in crops sown broadacre, in pastures and in weedy areas.

The number of sample quadrats will be determined by the pupal density and the desired sampling accuracy (see above). Unfortunately, pupal densities are frequently very low, less than 1 per row-meter and Wilson (1983) found 1 per 5 row-meters, thus a great deal of searching/sampling is required compared to that required where pupal densities are high, for example, 7 to 10 per row-meter.

1. Destructive Methods

Destructive sampling is probably the most accurate method used to estimate pupal density.

A. *Soil Sieving*

Soil samples can be sieved wet or dry, but pupae are not very robust, so many may be damaged by the sieving methods. (Soil cultivation is a means of reducing pupal survival.)

B. *Excavation*

Excavation of plots is less demanding of mechanical aids and has been used by several researchers (Caron, Bradley, Pleasants, Rabb, and Stinner, 1978; Lopez and Hartstack, 1985; Slosser, Philips, Herzog, and Reynolds,1975; Wilson, 1983). Emergence tunnels are first detected by scraping away the top 1 to 2 cm of surface soil with a flat-bladed shovel or trowel. Then the tunnels are excavated carefully to expose the pupal chambers. Pupae (or their remnants) are placed individually into specimen tubes using feather forceps and returned to the laboratory for identification to species and sexing. Some Tachinids, which emerge from the prepupal stage to pupate, can be collected from the *Heliothis* pupal chamber. Mortality factors (parasitism, disease, predation, etc) can be assigned with reasonable accuracy to most dead pupae.

The soil within each sampling area should be sorted carefully to a depth of about 10 cm to expose any pupae with tunnels not detected by surface scraping.

2. Nondestructive Methods

A. *Emergence Cages*

An alternative, nondestructive method for estimating pupal abundance is to measure moth emergence within caged areas (Caron et al., 1978; Roach, 1981; Lopez, Hartstack, and Beach, 1984). This method estimates total survivors (moths) rather than actual pupal numbers, but may be adequate for some studies.

3. Problems

Soil type and soil moisture determine the success of pupal sampling using the excavation method. In the brown and black cracking clay soils characteristic of the more important agricultural areas in Australia, the emergence tunnels are easy to expose if soil moisture is suitable. If too wet, then the soil is unworkable and sticks to the tools, too dry and the soils crumble and soil particles fall into and hide emergence tunnel openings. Suitable conditions for a particular soil type need to be determined by experimentation. In sandy soils, the emergence tunnels collapse easily and thus detection is more difficult, but this is compensated for by the relative ease with which sandy soils can be excavated and sorted to locate pupae.

With emergence cages, if pupal densities are low, then relatively large areas must be caged. However, large cages may be damaged by strong winds or hail, and microclimate and, consequently, insect development may be modified by a cage. Weather, predation, parasitoids, and pathogens all reduce moth emergence, but parasitism is the only mortality factor estimated by the cage method. Emergent parasitoids are not always a reliable estimate of parasitism because of the sometimes unexplained high mortality of pupal parasitoids (Wilson, 1983). Additional mortality can result when moths are trapped in their emergence tunnels and are unable to penetrate the surface soil. Destructive sampling techniques are necessary in order to quantify these specific mortality factors.

Chapter 6

Life Tables

P.M. Room, I.J. Titmarsh, and M.P. Zalucki

I. What Are Life Tables?

There are four processes that can change the size of a population: birth, death, immigration, and emigration. As originally developed for human demography, life tables were concerned only with rates of death and showed how life expectancy changed with age. In ecological studies, life tables can contain age-specific statistics for births, immigration, and emigration, but this is not appropriate for such animals as *Heliothis* spp. whose reproduction and long-distance dispersal occurs during one relatively short life stage. In these cases, single estimates of birth, immigration, and emigration rates for each generation provide quantitative links between successive generations.

Life tables are built to explain the observed numbers of individuals of different ages or life stages in a population. This is done by evaluating the importance (and usually the density-dependence) of different physical and biotic factors by correlation of the factors with age-specific death rates and population densities. It must be remembered that a factor may be correlated with an effect without having caused the effect!

The understanding gained from life tables can provide a rational basis for pest control by indicating which life stage should be attacked to have most effect in reducing numbers of the most damaging stage of a pest. Imagine, for example, that there was an average of three *H. armigera* eggs per cob in a crop of maize, that an ovicide was available that usually killed 50% of eggs, and that life table studies had shown that cannibalism among small larvae usually resulted in only one larva per cob surviving to a size at which significant damage was caused. Use of the ovicide would be irrational because the outcome would still be one damaging larva per cob, whereas the cost of the chemical and its application would be wasted as would be the pest-controlling effects of any beneficial insects killed by the chemical.

There are two conventional types of life table. A cohort (age-specific, horizontal) life table records age-specific death rates calculated from repeatedly counting a group of individuals born at the same time, as they get older. A static (time-specific, vertical) life table records age-specific death rates calculated from a single census of all the individuals of all ages present on a single occasion. Many ecology textbooks give examples of life table construction and analysis (Odum, 1971; Varley, Gradwell and Hassell, 1973; Southwood, 1978; Begon and Mortimer, 1981).

Both conventional types of life table can be built for *Heliothis* spp. but both have drawbacks. Cohort life tables require that particular individuals be counted repeatedly and this may be difficult or impossible without affecting the death rates which the study aims to measure (see Chapter 5). In addition, a cohort life table for individuals born at the start of the growing season of a host plant will not give a true picture of survival for individuals born towards the end of the season. Static life tables assume that generations are completely overlapping, having all age-classes present simultaneously, and that mortality factors have effects on particular age-classes that are constant through time. The first assumption may be correct for pests on some host plants, such as cotton, but false for pests of others such as maize that are suitable as hosts for a much shorter period. The second assumption must often be invalid because the size and quality of host plants and the abundance of natural enemies and alternative prey change so much with the time of year. For *Heliothis* spp., the cohort approach is likely to give more meaningful results than the static approach. In cases of crops that are attractive for long periods (for example, cotton), or crops that are attractive for shorter periods but planted at different times in different fields (for example, sunflower), several cohorts should be monitored starting at different times during the growing season.

A third, "rolling life table" approach that combines elements of both cohort and static life tables has been used for continuously overlapping generations of *Heliothis* spp. on cotton (Room, 1978). Numbers of all age classes were estimated every day in the manner used for static life tables, and temperatures were monitored continuously. Temperature records were used then with the known relationship of rate of development to temperature (Room, 1983) to calculate the proportions of cohorts each day that should have been present in each age class each subsequent day. Subtraction of the numbers observed in each age class from the numbers expected gave daily estimates of age-specific mortality. The main drawback of this method is the large cost of conducting daily censuses and processing (entering) the data.

II. Life Table Boundaries

It is important that the boundaries of a life table study be defined carefully because the statistics gathered refer to population performance averaged within the boundaries. For example, a life table for *H. armigera* based on lumped data from cotton and maize, or cotton varieties with and without nectaries, would not give information of much use for managing the pest in either crop or variety because of the different patterns of survival in each. Single life tables should be based on as specific a combination of crop variety, crop surroundings, soil type, climate, and time of year as possible. Comparison of life tables for different circumstances can give valuable insights for making such strategic decisions as which variety to grow and when to plant, as well as insights into the components of a cropping mosaic that contribute most to an area-wide pest problem. Where a factor seems to act in the same way in tables from different situations, results from single studies can be combined often to enable more rigorous statistical analysis.

The place where a study is conducted fixes many of the factors such as climate, soil type and surroundings, operating on a population. In contrast, the weather is so variable that results obtained in 1 year cannot be assumed to be representative for the same season at that place in other years. Weather occurring before and during a study will affect the results and it is recommended that life table studies for a particular place be carried out over a minimum of 3 years.

III. Data for Life Tables

Accurate determination of the age of individual *Heliothis* spp. in the field is impossible, thus developmental stages are used as a substitute. Larval instars can be separated but size-classes have been used often as an easier alternative (below). A system in common use identifies white (freshly laid) eggs, brown (developing) eggs, black (nearly ready to hatch) eggs, very small, small, medium, and large-sized larvae, pupae and adults.

Cohort life tables require individuals to be accounted for as they grow older, so the sampling interval should be no longer than the duration of a developmental stage. This varies with temperature and in practice inspections are usually carried out every 2 or 3 days. Eggs may be identified by marking the leaf to which they are attached, being careful not to touch them or mark too close in case the marker is toxic. Additional markers should be attached to the plant as larvae move about to help ensure that the correct individuals are found on the next occasion. Even so, larvae might be lost or confused with others not included in the study if they move significant distances between sampling occasions. Eggs and larvae can be labelled with

32P also and relocated using a Geiger counter (Room, 1977), though the procedure is more labor-intensive.

The same individual insects should be searched for on each subsequent sampling occasion and records made of the stage of development, or other fate if that can be determined by discovery of remains. Individuals cannot be followed through the pupal stage in the soil, so it is usual for larvae to be collected just prior to pupation and held in the laboratory to see which *Heliothis* spp. or parasites emerge or whether disease develops (see Chapters 2, 3, and 7). Losses of pupae to predation, unfavorable weather, or soil disturbance cannot be estimated when this is done.

Sample sizes for cohort life table studies are determined partly by the amount of mortality expected in the earlier developmental stages. Most mortality commonly occurs before larvae become medium-sized (Titmarsh, 1985) so that large numbers of younger stages must be followed to ensure sufficient data can be collected on older stages for statistical analysis. *Heliothis* spp. females lay between 1,000 and 2,000 eggs each and, as populations neither increase to infinity nor become extinct (except "locally"), each female must give rise to one female offspring on average. Hence, mortality between generations must be greater than 99% on average, and was found to be 95% between egg and pupal stages on tobacco (Table 6.1). This suggests that cohort studies should aim to start with thousands, rather than hundreds, of eggs if sufficient data on large larvae are to be collected.

Static and rolling life tables require single and repeated censuses, respectively, of all developmental stages. (See Chapter 5 for discussions of sampling techniques and numbers of samples needed.) The census data are needed for the construction of life tables and a second set of data, consisting of records of environmental factors, is needed for interpretation of the tables.

There are several general classes of mortality factors for pests including: insecticides, weather, predation, parasitism, disease, and poor nutritional quality of the host plant (that is, combinations of low concentrations of essential nutrients, low moisture content, and such defences as tough cuticle and toxic chemicals). It is probable that most mortality is caused by two or more of these factors acting together, such as disease, plant toxins, and inclement weather increasing susceptibility to predation, but such subtleties are unlikely to be disclosed by any but the most detailed and extensive life table studies. Where insecticides are used, they are likely to overwhelm and interfere with other causes of mortality whose effects should be measured on an unsprayed crop, nearby but sufficiently distant not to be affected by spraydrift.

Evidence of the ultimate cause of mortality rarely is observed directly, though empty egg shells and corpses bearing signs of attack by predators, and larvae in the terminal stages of diseases are sometimes found. In most cases, individuals simply disappear and the cause of mortality must be inferred from indirect evidence. If a series of life tables has been built, records of temperature, rainfall, humidity, growth stage, nutritional quality of the host, and numbers of predators allow regressions against numbers of individuals lost. Losses due to parasitism and disease can be estimated from samples of eggs, larvae, and pupae collected from near the study plot and reared in the laboratory (see Chapters 2 and 7). Experiments in which different permutations of predator species are excluded from or added to *Heliothis* spp. populations can supplement evidence obtained by regression (see Chapter 7).

A checklist of the categories of data that should be collected in a comprehensive life table study is as follows:

1. Field data: (a) *Heliothis* spp.: Numbers and, where possible, identity of white (fresh), brown and black eggs, very small (<3 mm), small (3–8 mm), medium (9–15 mm) and large (>15 mm) larvae, pupae, adults; (b) Insecticides: Types, concentrations, and times of application, if any; (c) Weather: Daily maximum and minimum air temperatures measured in a Stevenson's screen; these can be converted to temperatures experienced by different parts of the host plant (Room, 1983), humidity, rainfall, wind run, and pan evaporation; (d) Host plant quality: Plant growth stage and size, concentrations of nitrogen, toxins (for example, gossypol) and tannins; and (e) Predators: Numbers of immature and adult predacious Coleoptera, Heteroptera, Neuroptera, Hymenoptera, and Araneida.
2. Laboratory data: (a) Proportions of field-collected *Heliothis* spp. dying from particular diseases or parasitoids and, if healthy, giving rise to adult *H. armigera* and *H. punctigera*, and
3. Field experiment data: Survivorship after imposing such treatments as: (a) exclusion of all but one species or class (for example, flying/nonflying) of predator; (b) fertilization; (c) irrigation; (d) addition of competitors of the same, or different, species; and (e) application of chemical or microbial insecticides. At least three levels, or intensities of each treatment should be imposed with replication (see Chapter 7).

IV. Building Life Tables

Various degrees of detail may be shown in life tables (Begon and Mortimer, 1981), though the majority contain the following column headings:

x — developmental stage or age class;
a_x — number entering x;
$l_x - a_x$ transformed so that initial cohort size is 1,000;
d_xF — mortality factor responsible;
d_x — number dying during x;
$100q_x - d_x$ as a percentage of l_x;
S_x — survival rate within x;
$100r_x$ — real mortality, d_x, as a percentage of initial number;

Table 6.1. Cumulative Cohort Life Table for *Heliothis* Species on Vegetative Tobacco in Far North Queensland over Six Sites for the 1980–1981 Season[a]

x	a_x	d_xF	d_x	$100q_x$	S_x	$100r_x$	$100i_x$	k_x
Egg white	2,058	Dislodged	243					
		Desiccated	3					
		Infertile	5					
		Soil abrasion	4					
		Substrate dried	1					
		Unknown	10					
		Total	266	12.93	0.87	12.93	0.68	0.06
Egg brown	1,792	Dislodged	253					
		Desiccated	8					
		Infertile	21					
		Soil abrasion	9					
		Buried – cultural practices	3					
		Unknown	8					
		Total	302	16.85	0.83	14.67	0.93	0.08
Egg black	1,490	Dislodged	17					
		Desiccated	21					
		Soil abrasion	3					
		Buried – cultural practices	7					
		Unknown	4					
		Total	52	3.49	0.97	2.53	0.17	0.02
Larvae 1 + 2	1,438	Disappeared, had fed	377					
		Disappeared, had not fed	258					
		Disappeared, unknown if fed	27					
		Dead, desiccated, had fed	350					
		Dead, desiccated, had not fed	29					
		Dead, had fed	102					

		Mortality factor	No.					
		Dead, had not fed	45					
		Dead, unknown if fed	1					
		Drowned	6					
		Cannibalism	3					
		Hatching mortality	5					
		Predator—Mirid	1					
		Soil abrasion	12					
		Substrate dried	1					
		Buried—cultural practices	5					
		Unknown	16					
		Total	1238	86.09	0.14	60.16	28.28	0.85
		Disappeared	5					
		Dead, desiccated	2					
		Dead	2					
		Drowned	1					
		Buried—cultural practices	1					
Larva 3	200	Total	11	5.50	0.95	0.53	0.27	0.03
		Disappeared	8					
		Dead, desiccated	1					
		Cannibalism	2					
		Parasitized (*Microplitis*)	62					
		Unknown	1					
Larva 4	189	Total	74	39.15	0.61	3.60	2.94	0.21
		Disappeared	7					
		Dead, desiccated	1					
		Parasitized (*Microplitis*)	11					
Larva 5	115	Total	19	16.52	0.83	0.92	0.90	0.08
		Parasitized (*Microplitis*)	1					
		Parasitized (*Apanteles*)	1					
Larvae 6 +	96	Total	2	2.08	0.98	0.10	0.10	0.01

[a]Titmarsh, 1985.

$100i_x$—indispensable mortality that would not occur should the d_x under
   consideration be removed, as a percentage of initial number;
k_x—the difference between successive values of log l_x,
k_xF—k_x values partitioned among mortality factors acting in x; and
kappa—the partial sum of k values.

   In cohort studies, values of a_x are estimated by summing the numbers of
living and dead individuals found to have just entered each stage, x. In static
studies, values of a_x are estimated as the numbers of living individuals found
in each stage at the time of the single census. This can be fairly inaccurate
because, with totally overlapping generations, it assumes the equivalent of
zero mortality during the first half of the time spent in each stage. In rolling
studies with daily censuses, daily values of a_x are calculated from the numbers of individuals observed in an earlier stage and the duration of that stage
as determined by temperatures. For example, if ambient temperatures and
temperature/development relationships indicate that the small larval stage
should occupy 4 days, then one quarter of the small larvae found to be
present on day n would be expected to have entered the medium-sized larval
stage by day $n + 1$.

   Once a_x values have been estimated, manipulation of the data to complete
the construction of a life table is as indicated by the table headings. Mortalities (k_xF) can be shown linked to particular causes (d_xF) in cohort life tables
(Table 6.1) if evidence identifying the causes has been found. This cannot be
done in static or rolling life tables as the causes can be inferred only by
correlation of k_x values with records of weather, predators, and so on.

   Because the k_x and k_xF values are standardized to an initial 1,000 individuals, they may be compared between separate studies. Total mortality K is
given by the sum of all the k_x values, but in the case of *Heliothis* spp. this is
academic because no useful methods for estimating k values for adults have
been devised (but see Topper, 1987a). Until this problem is solved, complete
life tables to show how numbers in successive generations are related cannot
be built and kappa, the partial sum of k values, will be the best available
estimate of K.

V. Interpreting Life Tables

Mortality factors whose effects are seen in static and rolling life tables can be
identified by correlation only. A series of tables is accumulated with supporting records of environmental conditions immediately prior to each sampling occasion and correlations are found between particular k_x values and
environmental factors.

   The importance of mortality in particular developmental stages (k_x), or
due to particular factors (k_xF), in "regulating" population size can be determined by "key factor analyses" (Varley and Gradwell, 1960; Podoler and
Rogers, 1975). Regressions are performed of values of k_x or k_xF on K (kappa

for *Heliothis* spp.) from a series of life tables to identify the mortalities that contribute most to fluctuations in K (or kappa). A useful preliminary step is to plot partial and total mortalities as shown in Figure 6.1, to gain a visual impression of variance and correlation between mortalities. The degree to which particular mortalities are density-dependent can be determined in a similar way by regression of k_x or k_xF against a_x or l_x (Southwood, 1978), but see Royama (1981), Hassell (1985, 1987) and Hassell, Southwood, and Reader (1987) for discussions of difficulties in identifying mechanisms regulating population density.

In addition to mortalities, the number of individuals born into a population is likely to be significant in determining later population density. This can be tested for by performing regressions of the number of white eggs and various mortalities against numbers of large larvae or pupae and comparing correlation coefficients.

Computer software is available that can perform a number of useful analyses of cohort life tables such as assessing the similarity of survivorship curves (Hull and Nie, 1979).

VI. Testing Life Table Conclusions

Most conclusions derived from life tables should be regarded as hypotheses because they are based on correlations that might not reflect causal relationships. More confidence can be placed in such conclusions if attempts to falsify them fail. This can be done by analysis of independent sets of life table data or more directly by experimentation.

Laboratory experiments can show whether such factors as high temperatures or particular predators are capable of killing individual *Heliothis* spp., (Room, 1979; Qayyum and Zalucki, 1987), but they can indicate the numbers that would be killed in the field rarely. Field experiments can give a more realistic estimate, but it is usually impossible to exclude all causes of mortality other than the one under investigation and it is often difficult and expensive to replicate field experiments adequately. In any experimentation, it is crucial that the techniques used do not cause unwanted artifacts in the results. For example, a technique for excluding predators might prevent dispersal also — both of which might result in a higher population density.

VII. Using Life Table Conclusions

Conclusions derived from life tables become useful only when they are combined with other knowledge. For example, to understand region-wide population dynamics and changes in insecticide resistance, information is needed also on cropping sequences, the relative attractiveness of crops,

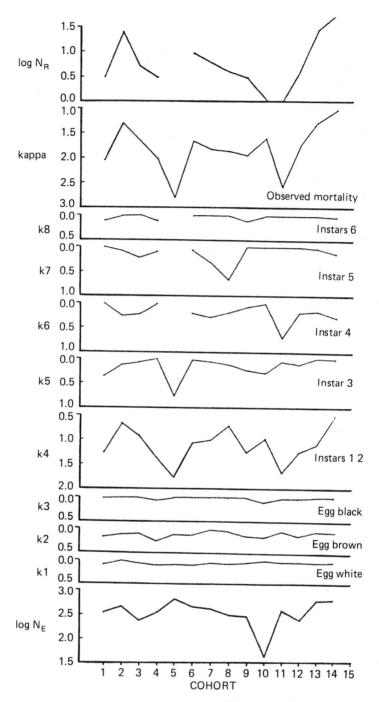

Figure 6.1. Variations in number of white eggs (N_E), k-values for eight age-classes, kappa, and number of larvae surviving to pupation (N_R) over 14 cohorts of *Heliothis* spp. monitored on tobacco in far north Queensland (the whole of Table 6.1 on pp. 74 and 75 describes 6 of the cohorts); Titmarsh, 1985.

prevailing winds, and insecticide use on particular crops. For management of individual crops, life table results need to be considered in conjunction with information on the life stage that causes most damage, the stage of the crop most susceptible to damage, and constraints on the use of control measures. In most cases, computer modelling is the only efficient way of handling the amounts of information involved and the often complex interactions between information of different kinds (see Chapter 14).

VIII. Examples of *Heliothis* Species Life Tables

Considering the large losses caused to a range of crops in Australia, India, Africa, Central America, and the USA, astonishingly few life table studies have been compiled for the species of *Heliothis*. Tables have been built for *H. armigera* on several crops in India (Bilapate et al., 1978, 1981, 1984), for *H. zea* on sweet corn in Hawaii (Vargas and Nishida, 1980), and for pooled *H. zea* and *H. virescens* on cotton in Mississippi (Hogg and Nordheim, 1983). In Australia, the only work has been the rolling life table study on cotton by Room (1978) and the cohort life table study on tobacco by Titmarsh (1985), of pooled *H. armigera* and *H. punctigera* in both cases.

None of these life tables is complete because all omit the adult stage and estimates of the potential number of eggs that could have been laid but were not [identified as a key factor for a number of Lepidoptera by Dempster (1983)]. With the notable exception of Titmarsh (1985), none of the studies attempted to assess the importance of all causes of mortality affecting immature stages, focusing instead on particular causes such as parasitism. Most of the life tables showed that most mortality occurred in the egg and early larval instar stages, but the results that can be compared meaningfully are far too few to even hint at which factors might regulate *Heliothis* spp. populations commonly.

Chapter 7

Evaluating Natural Enemy Impact on *Heliothis*

J.E. Seymour and R.E. Jones

I. Introduction

Natural enemies of *Heliothis*, like those of other insects, fall into three categories namely, pathogens, parasitoids, and predators, that require very different techniques for their study. Pathogens are briefly discussed elsewhere (Chapter 2 and 3) and will not be covered here.

In considering organisms as potential biological control agents, the conventional wisdom has suggested that the best agents will have the following properties:

1. They will be host-specific.
2. The seasonal timing of their activity will be synchronous with that of the pest.
3. They will exhibit a rapid numerical response, that is, they will increase in density rapidly when host densities increase.
4. Each individual will need only a few host individuals (preferably one) to complete its life cycle so that populations can persist when prey populations have been driven to a very low density.
5. They will have very high searching abilities.

Particularly because of criteria (1) and (2), parasitoids have been regarded as the best candidates for natural enemy introduction programs, and it is clear that a large number (though regrettably, a small proportion) of such introductions have been successful. As Murdoch, Chesson, and Chesson (1985) cogently point out, the evidence on which this belief is based is at best equivocal: it is certainly possible to make a compelling case for the superior effectiveness of general predators, that do not require pest populations to be present in order to maintain themselves, but that can have a major impact on them when they do appear.

The argument about which sort of control agent is most likely to be effective is not necessarily of prime importance when practical decisions about biocontrol introduction have to be made. Even if we were to conclude that general predators were potentially more effective, the fact that a specialist parasitoid can be (more-or-less) guaranteed not to have undesirable side effects may lead us to prefer it. However, if our control strategies include the possibility of enhancing existing natural enemies rather than introducing new ones, the argument becomes very important indeed. At present, we do not have the data to resolve it.

In the case of *Heliothis*, especially in Australia, we have almost no substantive information about the impact of natural enemies in any cropping system for any major region. The only detailed life table studies of *Heliothis* species in Australia cover populations attacking tobacco in North Queensland (Titmarsh, 1985), a system in which plant-related factors appear to act as the primary mortality agents. In Australia, as elsewhere, a large number of natural enemies are known to attack *Heliothis*. Room (1979) recorded 4 species of Dipteran and 12 species of Hymenopteran parasitoids from *Heliothis* spp. on cotton in the Namoi valley, and found evidence that at least 19 species of insect, 5 species of spider, and house mice, were predators. Other workers have documented the presence of a variety of potential predators and parasites in other parts of Australia and on other crops (Wright and Nitkin, 1964; Twine, 1973; Bishop and Blood, 1980; Broadley, 1981 a,b; Forrester, 1981; Pyke, 1981). Samson and Blood (1980) provided evidence for the potential impact of 3 predators based on laboratory studies. Bishop and Blood (1981) correlated spider abundance with *Heliothis* abundance in southeast Queensland cotton fields, but did not attempt to evaluate the impact of spider predation (though they were left with strong intuitive impressions of the magnitude of the impact). In 1980, the Queensland Department of Primary Industry held a workshop on the biological control of *Heliothis* at which a number of contributors documented the presence and in some cases the abundance, of potential or actual natural enemies in a range of cropping systems. Nobody at that workshop presented any data to indicate the impact of natural enemies in their cropping system, except for the work by Titmarsh (1985) mentioned earlier, and a brief report of the results of an experimental release of Trichogramma in a tomato crop (McLaren, 1981). When a more general workshop on *Heliothis* ecology was held 5 years later, it became very apparent that this situation had not changed. Only one of the submitted papers dealt with natural enemy impact, and that one (Murray and Wicks, 1986) pointed out that braconid parasitoids in the genus *Microplitis* (now reassigned to *Microgaster*) might be having a very substantial impact in some systems, and warranted more attention.

It will be evident from this brief review that knowledge of natural enemy impact on Australian *Heliothis* populations is somewhat incomplete! This deficiency is certainly not unique to the Australian situation or to *Heliothis*. Although it is certainly possible to find a large number of papers whose titles are something like *The impact of predator or parasite X on prey Y*, it is much less common for such papers, read with proper textual attention, to live up to the promise of their titles. There are some exceptions: perhaps the most notable involving *Heliothis* is a North American study by Hogg and Nordheim (1983), that was able to attribute a significant part of the observed consistent seasonal differences in juvenile *Heliothis* survivorship to the action of natural enemies.

One need not seek far to find the reason for this deficiency. In general, evaluating natural enemy impact is a problem of considerable technical difficulty. The impact of parasitoids is perhaps somewhat easier to determine, but is by no means free of traps. The impact of general predators is probably impossible to evaluate without highly labor-intensive procedures or some form of carefully considered experimental intervention. For reasons considered later, this applies especially to the suites of predators common in many Australian ecosystems.

To evaluate the impact of parasitoids and predators, we need to have some way of estimating what the abundance or survivorship of prey would have been in the absence of natural enemies. For this reason, the most convincing studies of natural enemy impact, on *Heliothis* or anything else, generally have been done in combination with studies of the pests' population dynamics, and especially in conjunction with life table studies. Methods of estimating the abundance of *Heliothis* populations (Chapter 5), and of developing life tables for it (Chapter 6), are discussed elsewhere and hence, need not be considered here. However, we do consider briefly methods of estimating the densities of natural enemy populations.

Because the necessary methodologies are so different, first we consider the study of parasitoids, and then proceed to an examination of techniques for studying predators.

II. Parasitoids

1. Techniques for Estimating Abundance and Attack Rates

For parasitoids, the difficult task of estimating the abundance of adult stages is undertaken rarely, though there are exceptions! Lingren, Lukefahr, Diaz, and Hartstack (1978a) counted cocoons and adult parasitoids present on plants; Powell and King (1984) used sticky traps with a virgin female as bait to catch males. Usually, samples of eggs and/or larvae are collected and returned to the laboratory until emergence of the parasitoids, thereby providing the data for estimates of juvenile parasitoid abundance and age structure, (though it is rare for these parameters actually to be calculated) and estimates of parasitism rates for eggs or larvae of the host population. (For examples involving *Heliothis*, see Lewis and Brazzel, 1968; Lewis, Sparks, Jones, and Barnes, 1972; Lingren et al., 1978a; Burleigh and Farmer, 1978; Oatman and Platner, 1978; Johnson, 1979; Room, 1979; Broadley, 1984; Stadelbacher, Powell, and King, 1984; Johnston, 1985; Puterka, Slosser, and Price, 1985; Titmarsh, 1985.)

Thus, collecting the necessary field data is in principle a relatively straightforward procedure. Collecting an unbiased sample of parasitized and unparasitized larvae may be difficult if parasitized larvae are distributed differently on the plant from unparasitized larvae. Certainly, this phenomenon has been recorded in other lepidopteran species (Stamp, 1982); it does not seem to have been recorded for *Heliothis*, but as far as we are aware, nor has it been looked for yet. The simplest method of avoiding biased collections from this source is to search for and collect all the larvae on each plant examined. This also solves the problem of unequal apparency that may occur when parasitized larvae are smaller and less mobile than unparasitized larvae of the same instar, as is the case for *Heliothis* larvae parasitized by *Microgaster* (Seymour, 1985).

A further complication in the case of *Heliothis* spp. larval parasites arises from the difficulty of reliably separating the larvae of *H. armigera* and *H. punctigera*. In many of the studies cited earlier, the larvae have been lumped together with no distinction being made between species. Alternatively, the proportions of each species for parasitized larvae are assumed to be the same as for unparasitized larvae collected simultaneously and reared to adulthood; that is, parasitoids are assumed to attack both species with equal likelihood. However, this is a dangerous assumption: Seymour (1985) used laboratory preference trials to demonstrate the existence of strong locality-specific preferences by *Microgaster demolitor* for different *Heliothis* species. This reinforces the field data of Titmarsh (1985), who suggested that Mareeba populations may only attack *H. armigera*. This problem would disappear if we had a reliable *Heliothis* larval key (see Chapter 1, 2, and 13).

In order to obtain reliable estimates of net larval parasitism rates, it is important to record the stage of the host larva at the time of collection as precisely as possible (that is, preferably to record the instar rather than simply grouping larvae as "small", "medium", and "large"). *Net larval parasitism rate* is here used to mean the proportion of newly hatched larvae that, in their subsequent development, are killed by parasitoids. This will not correspond to the estimate of gross parasitism rate (that is, total number of parasitized larvae divided by the total number of larvae), unless parasitoid attack occurs immediately after hatching, the parasitoids do not alter the developmental period of the larvae, and the parasitoids emerge immediately before the larva pupates. In fact, no parasitoid does behave in this way, so that to evaluate the true impact of parasitism, all deviations from this "ideal" behavior must be allowed for. Appropriate estimation methodologies for *Heliothis* larval parasites are discussed in detail by Titmarsh (1985). The simplest procedure is to use only larvae at a stage close to the end of the period when they are vulnerable to parasitoid attack, but before the juvenile parasitoids are likely to have completed their development and emerged. Titmarsh's (1985) results suggest that the fourth instar may be the most appropriate stage to use to evaluate parasitism rates of the braconid *Microgaster*, the fifth instar for *Heteropelma*, and for tachinids, the sixth. Instar preference trials carried out in the laboratory for *Microgaster* by Seymour (1985) confirm that this species prefers second and third instar larvae for oviposition, and is rarely able to attack the oldest instars without sustaining physical damage.

If the parasitoid extends the length of time spent in the instar from which estimates are made (as is usually the case), then raw parasitism rates will overestimate net parasitism rates, since parasitized individuals remain available for collection for a longer period. The appropriate correction, if the relative development times of the parasitized and unparasitized instar are known, is:

$$\text{Corrected parasitism rate} = 1 - (1 + k)S/2$$

where k is the ratio (development time of parasitized instar)/(development time of unparasitized instar), and S is the proportion of unparasitized larvae in the sample. (This formula also provides an appropriate correction in the less likely case of parasitism-shortening development time of an instar.)

The problems described above are bypassed if parasitization rates are obtained as a byproduct of a life table study of juvenile *Heliothis*, in which the fates of individual eggs and larvae are followed through their lifespan. This allows parasitism rates to be obtained directly, without the need for application of any corrections. However, there must be enough survivors in the monitored cohort to provide good estimates of the time when parasitoids emerge!

A further problem, and one less easily dealt with, occurs if the mortality rates of parasitized and unparasitized larvae differ. Jones (1987) demonstrated that parasitized larvae of *Pieris rapae* were substantially more susceptible to ant predation than were unparasitized larvae. In life table studies of *P. rapae*, this resulted in an inverse correlation between larval survival rate and the parasitism rate among the surviving older larvae (Jones, Nealis, Ives, and Scheermeyer, 1987). Some anomalous patterns in Titmarsh's (1985) data suggest strongly that a similar phenomenon may have occurred in *Heliothis* populations on tobacco at Mareeba. If so, then rates of parasitism in collected larvae, in fact may underestimate the impact of the parasitoids substantially. It would be of some value to confirm or refute this possibility.

The same problems of obtaining corrected estimates apply to egg parasitism as to larvae, that is, errors in the estimate of net egg parasitism rate from egg collections are introduced, if the act of parasitism occurs some way into the egg developmental period, and if the parasitized egg remains on the plant for longer than an unparasitized egg would have done. Because the egg may not have identifiable stages of differential susceptibility, and until better information is available, it is probably best to assume that parasitism occurred midway through the developmental period if gross parasitism rates are less than 50%, and earlier if they are higher.

III. Predators

1. Techniques for Estimating Abundance

Because predator species include such a diversity of animal groups, techniques for sampling the potential predator community are unlikely to sample all components of the fauna with equal efficiency. In particular, if predation by vertebrates is important, completely independent estimation techniques will be needed. Here we restrict attention to arthropod predators.

One of the more regrettable characteristics of sampling methods for terrestrial arthropods is that, in general, there is a strong inverse relationship between ease and precision of sampling. Methods intended to estimate "relative abundance", usually by some form of catch-per-unit-effort method, may be both fast and simple, but are, in most cases, highly unreliable (see Chapter 5). This was demonstrated graphically by a study of sampling techniques in cotton carried out by Byerly, Gutierrez, Jones, and Luck (1978). Each week for the 4-month growing season, these workers compared the two most commonly used relative sampling methods — sweep-netting and D-Vac sampling — with a much more laborious technique involving a total count of all the arthropods on a sample of cotton plants that had been bagged and cut very early in the morning (the time was chosen to coincide with minimum activity levels and flight probabilities for arthropods resident on the plants). Some of the predator species present in substantial numbers were barely detected by the two relative methods, and even for species appearing abundantly in samples collected by the relative methods, those methods usually did not even detect general population trends.

Byerly et al. (1978) concluded that for all but a few species, it was not possible to "calibrate" the relative methods to generate reliable abundance estimates.

Southwood (1978) discusses in detail the commonly used methodologies for estimating insect abundance, including mark-recapture techniques, that may be quite successful for some predator populations, and methods used to sample a range of different arthropods from various kinds of plants. The growth form of the plant may often dictate an appropriate sampling method. Because *Heliothis* exploits such a wide range of plants, the range of sampling techniques needed for the associated fauna may be correspondingly variable. Of course, the purely statistical problems involved in developing an appropriate sampling design apply just as strongly to predator populations as they do to *Heliothis* itself (see Chapter 5).

2. Techniques for Estimating Consumption Rates

If we have some estimate of the abundance of a particular predator, then in order to evaluate its impact, we also need to know how many *Heliothis* each can be expected to eat, and how old the prey are expected to be. Laboratory trials of consumption rates of particular predators usually tell us only how much prey the predator will eat if it has no choice and if the problems of locating the prey are solved for it. For a set of predators on cotton plants in California, R.E. Jones et al. (unpublished) found that consumption rates obtained in laboratory trials were highly misleading. Some predators that appeared voracious in petri dishes (a hemipteran, *Geocoris pallens* and a beetle, *Collops vittatus*) proved to be ineffectual in the field: *G. pallens* because it stopped searching at high temperatures, and *C. vittatus* because, given the choice, it was more interested in pollen than in prey. Thus studies intended to evaluate the voraciousness of particular predators should be undertaken in conditions as realistic as possible – and preferably in the field. There are a variety of methods available to estimate consumption rates in field conditions, none of them is entirely satisfactory.

A. Direct Observation

Perhaps the most straightforward way to evaluate the predation rates of particular predators is simply to sit in the field and record the behavior of individual predators for long periods. This method is convincing, free of most of the sources of error of other methods, and, especially if prey densities are also known, may be very illuminating (the study of Californian cotton predators mentioned above was carried out in this way). It is also exceptionally tedious, as predatory insects, like other insects, spend a large part of their time doing nothing at all. Moreover, even when the predators are active, large amounts of observation time are needed to generate adequate estimates of predation rates. Direct observation has been used very successfully in predator-prey studies involving prey species other than *Heliothis* (Frazer and Gilbert, 1976). However, few workers have used it if they could think of a plausible alternative.

B. Examination of Gut Contents

For larger predators, especially vertebrates, remnants of insect prey may be visible or identifiable in the gut contents and in faecal material. Room (1979) discovered that house mice consumed *Heliothis* larvae and pupae by an examination of their gut contents. If we also know how long a particular food item may be expected to remain in the gut, then this could provide estimates of consumption rates.

C. Radioisotope Labelling of Prey

This technique was also used by Room (1977, 1979), to identify predators. He placed eggs or larvae labelled with ^{32}P on cotton plants and tested possible predators for radioactivity when caught in the same area 24 h later. Although Room used both gut content analysis and the labelling technique simply to survey and identify predators, both techniques have been used by other workers to estimate consumption rates by the predators being examined. However, like precipitin tests, radioisotope labelling may give misleading results if used to obtain quantitative estimates of the numbers of prey attacked. The problems are discussed in more detail below.

D. Precipitin Tests

In this test, described by a number of workers (Dempster, 1960; Loughton, Derry, and West, 1963; Rothschild, 1966; Ashby, 1974), the specific proteins of the prey are identified by their reaction with the serum of a sensitised mammal (usually a rabbit). A list of variations in test procedures is given by Southwood (1978). If some assumptions are made about the distribution of prey captures among predators that test positively, numbers of prey consumed can be estimated. There are a number of problems with the technique, which are also identified by Southwood: cross-reactions between closely related prey species are common, not all stages of the prey may react equally well to the antisera, and the prey may have been already dead when eaten. A further problem of particular relevance in Australian ecosystems, is that results may be misleading if the predator is a social insect: in particular, an ant. In this case, many different predator individuals may eat part of the same individual prey, and quantitative estimates obtained from precipitin tests that equate the number of prey attacked to the number of predators that test positively, are likely to be highly misleading.

E. Exclusion Experiments

The impact of predation may be demonstrated by comparing the survival rates of eggs or larvae on plants from which the predators have been excluded, with those on plants to which they have free access. A variety of exclusion techniques are available; they usually involve either caging the plant or, for nonflying predators, using some sort of barrier to limit access to the stem. Cage exclusion is considered as part of a larger class of cage experiments in the next section; barrier methods are discussed here briefly. The problem with exclusion experiments is that, although they may demonstrate that predation has a substantial impact, they may not be able to determine which predators are responsible for it. Sticky barriers on the stem of a plant, for example, will exclude ants, earwigs, ground beetles, some spiders, and a variety of other potential predators. Sometimes a combination of exclusion and low-intensity observation will provide more precision. Jones (1987) was able to assign the effects of predation on late-stage *P. rapae* larvae to the activity of a particular species of ant, using a combination of exclusion experiments and daily monitoring of the ant species present on each plant.

F. Cage Experiments

Cage experiments, in which known numbers or densities of prey are exposed to known numbers of particular natural enemies, are potentially very sensitive methods for evaluating natural enemy impact. However, it is necessary to bear in mind that what is being evaluated is the impact of predators in cage conditions—the relevance of the results to the situation in the field will depend on the similarity of conditions in the cage to conditions in the field. As noted earlier, trials carried out in Petri dishes, which can be regarded as very small and un-field-like cages, may provide very misleading results indeed. Beginning with the classic experiments of Huffaker (1958), numerous studies have demonstrated the importance of the complexity of the habitat within which the predator must locate its prey: in general, more complex structures reduce predator efficacy. Cage experiments, therefore, need to provide a level of environmental complexity comparable with the relevant field conditions. Also it may be important that the physical environment mimic field conditions at the time and place of interest. For example, because predator and prey are both poikilotherms, and do not necessarily share the same temperature requirements, predator impact may vary markedly with ambient temperature, even when the time scale is the physiological time of the prey. A commonly observed pattern in temperate climates is for the predator to have higher temperature requirements than the prey, and consequently to have less impact at low temperatures.

Most of the procedures discussed above require the design of field experiments within which they may be applied. Critical surveys of the ecological literature (Hurlbert, 1984; Underwood, 1986) demonstrate that it is not at all easy to design a good field experiment, and disastrously easy to design a bad one. As Connell (1983) points out, the possession of data from a badly designed experiment is often worse than having no information at all; it is worth taking considerable pains and obtaining appropriate statistical advice over experimental design.

IV. The Context of Natural Enemy Studies

The methods described above offer ways of assessing the proportion of *Heliothis* eggs or larvae killed by predators or parasitoids, and in conjunction with *Heliothis* life table studies, can tell us whether the impact of natural enemies is sufficient to produce a reduction in population numbers over the generation that has been studied. The problem with such studies, of course, is that even if we restrict our attention to a particular locality and a particular cropping system, doing only one such assessment is pointless. The answer obtained is almost certain to change with season, weather, cropping stage, prey density, and a variety of other factors. Natural enemies may show a numerical response that results in very low early-season densities, may have diapause (Kay, 1982a) or migratory patterns that do not match those of *Heliothis*, are almost certain to have different temperature preferenda, and, for any of a host of other reasons, may be unavailable or ineffective for part of the time that *Heliothis* spp. present problems. Can natural enemies control *Heliothis*? is the wrong question. The right question is When can natural enemies control *Heliothis*? The challenge is to design studies comprehensive enough to answer it!

Chapter 8

Measuring Development of *Heliothis* Species

P.G. Allsopp, G.J. Daglish, M.F.J. Taylor, and P.C. Gregg

I. Introduction

The "life history" approach to population ecology uses the interactions of biological factors with each other and components of the environment of the subject organism to explain population performance (Kitching, 1977, and Chapter 14). In insects, as in all poikilotherms, the interactions are complicated by a temperature-mediated "time scale". This time scale is the fundamental driving variable in the population dynamics of insects. It features prominently in all population dynamics models of *Heliothis* spp., for example, MOTHZV-2 (Hartstack, Witz, Hollingsworth, Ridgeway, and Hunt, 1976), and in models used to aid in *Heliothis* management, for example, SIRATAC (Hearn, Ives, Room, Thompson, and Wilson, 1981) (see Chapter 14).

In this chapter we deal with the influence of temperature on development and ways of measuring and modelling these effects. Differences in microhabitat temperature, behavioral thermoregulation and temperature fluctuation, host plant effects and intraspecific variability are discussed. Previous temperature-related studies with Australian *Heliothis* spp. are reviewed.

II. Temperature-Development Relationships

Insect development occurs within a definite temperature range. If development times are measured over that range, and these measurements are taken at close temperature intervals, a backwards "J"-shaped curve results (Fig. 8.1). If the reciprocals of development times (= rates) are plotted against

Figure 8.1. Typical relationship between development time and temperature at constant temperatures.

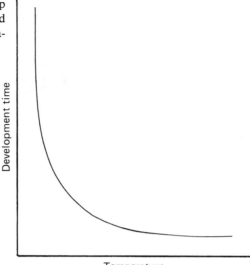

Temperature

temperature, a shallow sigmoid curve results (Fig. 8.2). At the lower thermal limit (region B of Fig. 8.2), the development-rate curve asymptotically approaches the point of zero development. Insects often survive relatively long periods at low temperatures with little or no development. Hence, the temperature at which development first occurs, the threshold, is difficult to measure accurately. As temperature increases, development rates become proportional to temperature, resulting in an approximately linear region in the middle of the response curve (region A of Fig. 8.2). Development then begins to slow, reaching a maximum at the so-called optimal temperature. Thereafter, development falls off sharply (region C of Fig. 8.2). Mortality rates are often high in this region, making the study of development difficult.

Many empirical and biophysical models describe the time versus temperature, or the rate versus temperature curves of insect development, and are used widely to predict insect development times. Below we describe some of these models. Most are based on development under constant temperatures — an artificial situation for most insects.

There has been some use of fluctuating temperatures. Most of the models are deterministic, although Wagner, Sharpe, and Coulson (1984a) and Dennis, Kemp, and Beckwith (1986) have both described methods for incorporating variability.

1. Linear Model

The most widely used approach to predicting insect development uses a linear approximation of the rate versus temperature curve. The products of

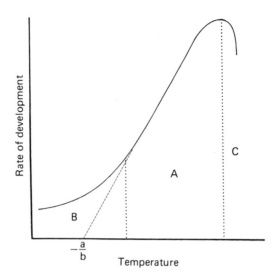

Figure 8.2. Typical relationship between development rate and constant temperatures. The curve can be divided into three regions. Region A, The approximately linear region where the degree-day model ($y = a + bT$) often gives satisfactory results, and where the error due to the effect of nonlinearity on rate summing is minimal; Region B, Where development rate asymptotically approaches zero rather than the threshold temperature estimate ($-a/b$). Here the number of degree-days per unit time will be underestimated, and errors associated with the effect of nonlinearity on rate summing will overestimate the time required for development also; Region C, Development rate declines after reaching an optimum. The decline may or may not be symmetrical about the optimum. In this zone, the linear model will overestimate the number of degree-days required for development and rate summing errors will underestimate the time required for development.

development time and the number of degrees above a threshold temperature are assumed constant. This leads to the so-called thermal summation principle. The duration of development is calculated by adding up the number of thermal units (degree-days or degree-hours) contributed at each temperature. The method is used widely because it requires minimal data for formulation, is easy to calculate, and often yields approximately correct values [Gregg (1982) and Allsopp (1986)]. The lower temperature threshold is determined empirically by extrapolating the intermediate, approximately linear, part of the curve. Thus, the approach is valid only where temperatures in region A of Figure 8.2 are used and the number of degree-days required for development is underestimated near the lower threshold and overestimated near the optimum temperature. The latter problem can be reduced by incorporating an upper threshold, above which the number of degree-days is held constant (Fig. 8.2). Standard errors for the lower threshold temperature and the thermal constant (reciprocal of the slope of the line) can be calculated using the methods of Campbell, Frazier, Gilbert, Gutierrez, and Mackauer (1974).

Higley, Pedigo, and Ostlier (1986) contend that degree-day accumulations are influenced by eight factors: (1) substrate availability (influenced by nutritional factors and temperature fluctuation); (2) enzyme availability (influenced by hormonal factors and temperature fluctuation); (3) approximations and assumptions in laboratory estimates of development; (4) approximations in calculating the developmental minimum; (5) approximations in calculating (or not calculating) the developmental maximum: (6) approximations from using single values for thresholds rather than thresholds that change with age; (7) behavioral thermoregulation; and (8) propriety and limitations of temperature data used in calculations. Some of the difficulties caused by these factors can be overcome by the use of the nonlinear models described below. Others are more fundamental and will be discussed below as sources of variation.

2. Sigmoid Function

Davidson (1942) proposed a logistic equation of the form

$$y = \frac{A}{1 + e^{(B + Cx)}}$$

where y = development rate, x = temperature and A, B, C are constants. This equation is not very descriptive at one or both ends of the response curve. This can be partially overcome by the relationship of Stinner, Gutierrez, and Butler (1974a) which is similar except that it decreases at temperatures above the optimum (z) — for $x > z$, $x' = 2z - x$. The curve is symmetrical around the optimum and for this reason has drawbacks at higher temperatures.

3. Pradhan Function

Pradhan (1946) used an exponential equation similar to a normal distribution of the form

$$y = Ae^{B(C-x)^2}$$

Like the Stinner function, the curve is symmetrical about the optimum.

4. Logan Functions

Logan, Wollkind, Hoyt, and Tanigoshi (1976) attempted to overcome the problem of symmetry about the optimum by combining two exponential equations to describe temperatures above and below the optimum. The uniformly valid additive composite is given by

$$y = A(e^{(Bx)} - e^{(BC-D)}), \text{ where } D = (C - x)/E$$

where A, B, C, and E are constants. They noted that the development response in the low-to-mid-temperature range was more sigmoid than exponential, and presented another model

$$y = A\{[1 + Be^{(-Cx)}]^{-1} - e^{(-D)}\}, \text{ where } D = (E - x)/F$$

where A, B, C, E, and F are constants. Both models are more descriptive than either the Stinner or Pradhan models, and the five-parameter model is slightly more descriptive than the four-parameter model. Both models require a number of data points above the optimum, data which most workers have found difficult to acquire.

5. Polynomial Functions

Authors such as Tanigoshi and Browne (1978) and Harcourt and Yee (1982) have used polynomial functions to predict development times of insects. Although these functions have no biological basis, there are several advantages in their use: parameters can be estimated easily using leastsquares estimation; a confidence interval can be determined for each parameter; the fit of the model can be improved readily by the addition of high-order terms; and the polynomial is not necessarily symmetrical about an optimum. For empirical predictions this asymmetry is, however, acceptable.

6. Biophysical Model

Sharpe and DeMichele (1977) consolidated the work of a number of earlier workers to formulate a complex biophysical model that describes the nonlinear response in development rates at both high and low temperatures, as well as the linear response at intermediate temperatures. The model describes the entire response curve over a restricted range. Wagner et al. (1984b) have provided a computer program that makes parameter estimation easier. The biophysical model is based on enzyme kinetics and claims have been made for a theoretical justification that is not shared by the other nonlinear models. This claim is only warranted if there are no fluctuating temperature effects of the kind described in the next section.

7. Maximum Likelihood Method

Tokeshi (1985) proposed a method using a maximum likelihood method for estimating the minimum threshold temperature and degree-days required to complete growth. Although he used a linear relationship, there is no reason why the relationship must intrinsically be linear. The method can be used equally well with fluctuating and constant temperatures.

8. DEVAR

Dallwitz and Higgins (1978) developed the computer program DEVAR that allows development-temperature relationships to be determined from rearing under fluctuating temperatures. Any effect of temperature fluctuation are incorporated thereby in the model. The program is sufficiently robust to allow the use of any type of development rate-temperature relationship. Allsopp (1986) used DEVAR to construct models for development of false wireworms under fluctuating temperatures and found that these relationships modelled development better than did relationships derived from rearing under constant temperatures.

9. Podolsky Phenocurves

Podolsky (1984) presented a system for modelling temperature development using two types of empirical equations. One deals with the temperature characteristics of a geographical region. The second relates the length of a particular stage of an organism's development to the mean temperature during this period through a phenological curve or phenocurve. These phenocurves allow development under fluctuating conditions to be quantified and the relationship can take any mathematical form. Because minimal equipment is required to collect data, Podolsky (1984) maintained that the method is simple to use and has wide applicability. However, Allsopp (1988) showed that Podolsky models were inferior to those derived with DEVAR for modelling development of false wireworms in fluctuating temperatures.

III. General Experimental Requirements

Howe (1967) reviewed the requirements for valid temperature-rate studies, with particular reference to insect pests of stored products. Many of his conclusions are also relevant for *Heliothis* spp.

1. Number of Temperature Treatments

It is possible to derive an approximate degree-day relationship with as few as three or four temperature treatments if they fall within the intermediate zone of Figure 8.1. To describe a nonlinear curve, at least 10 temperatures separated by intervals of no more than 2.5°C are required (Howe, 1967). To distinguish between alternative nonlinear models even more temperatures may be required. In particular, to distinguish between models that are symmetrical about the optimum and those that are not, several temperatures at intervals of about 1°C around the optimum are needed.

Few experiments with *Heliothis* spp. meet these stringent requirements. For the purposes of pest management this may not matter, because functional models can be devised without meeting all criteria. However, it is unwise to compare the merits of different models, and even more so to speculate on their physiological implications, unless data of this quality are available.

2. Replication

Intraspecific variation in the response of insects to temperature may be considerable, so extensive replication is advisable. At least 30 or 40 insects surviving to the end point in each temperature treatment should be the aim. At temperatures near the upper and lower limits, mortality may be high and allowance should be made for this. This mortality may not be random with respect to individual variation in developmental rate. In the absence of data on any such biases, increasing the sample size will probably be appropriate. Mortality rates at each temperature should be reported always so that the possibility of bias can be assessed.

3. Temperature Control

With the type of equipment used in most entomological laboratories it is quite difficult to measure, much less to control, temperature to an accuracy of better than 0.5°C. Errors of more than this can spoil the fit of models and make it impossible to distinguish between them (Browning, 1952). Diurnal fluctuations and spatial variation within incubators are possible sources of error, as is temperature change due to radiant energy from lights. The problems are magnified when controlled fluctuating temperatures are required. These difficulties suggest that where possible, replication of temperature treatments in different incubators is a wise precaution. They also provide further reasons for caution in comparing models.

4. Frequency of Observation

It is necessary to compromise between the labor requirements of frequent observation and the loss of precision if observations are too infrequent. Howe (1967) recommends observations in a range of at least 10 of the time units around each experimental end point. Fewer than five indicates that the variance is not worth estimating and the mean is inaccurate. Such automated methods as the use of automatic cameras or emergence detectors connected to clocks or data loggers are ideal (Gregg, 1981).

5. Statistical Description of the Development Time

Development times may be described by three methods. The time to emergence of the first individual can be used, but it is generally too dependant on sample size and too vulnerable to contamination of the sample by an older

individual. The mean and the median development times are better alternatives. Messenger and Flitters (1958) argue in favor of the median on the
grounds that it is less affected by abnormal values than the mean.

IV. Sources of Variation

1. Field Temperatures

The major problem in predicting the development of field populations of
insects from laboratory development-temperature data is the relative heterogeneity of microclimate in the field. Whereas the temperature in a laboratory incubator cabinet with uniform artificial illumination is very close to the
temperature experienced by the insect, temperatures in the microhabitat of
the plant or soil may depart widely from the ambient (shade) temperatures
recorded by the local meteorological station. This problem is reviewed more
fully by Willmer (1982). There are three ways of correcting the problem: (1)
determine development-temperature relationships under field conditions;
(2) relate meteorological station temperatures to microhabitat temperatures
by predictive equations; and (3) measure microhabitat temperatures directly.

Prediction of microhabitat temperatures from station temperatures is likely to be most precise for uniform crop environments. Fye (1971) and Room
(1983) have fitted such predictive equations for cotton canopy temperatures.
Individual crops can be expected to depart from such predictions due to
local soil conditions, albedo, wind patterns, and crop phenology. Direct
measurement of microhabitat temperatures is a better alternative. Small
low-cost solid-state microprocessors are now available that allow microhabitat temperatures, as well as humidity, solar radiation, and rainfall
to be recorded from many stations (Maywald, O'Neill, Taylor, and Baillie,
1985).

2. Behavioral Thermoregulation

Behavioral thermoregulation by insects (alternate sun basking and shade
seeking) is a further source of error in predicting development rate, even if
microhabitat temperatures are used. The magnitude of such variation may
be significant for some species. Gregg (1981) found that models for development of the Australian plague locust, *Chortoicetes terminifera*, that did not
include terms to account for behavioral thermoregulation underestimated
development rates by up to 35%. Body temperatures could exceed those of
the ambient air by up to 15°C. In the case of *Heliothis* spp., the effects of
behavioral thermoregulation are unknown, but observations of larval behavior in the field (Mabbett et al., 1980) suggest that they may be important
enough to warrant consideration.

In laboratory experiments, radiation is usually uniform and weak relative to sunlight, so models based on laboratory data will not take account of behavioral effects. The problem can be overcome by measuring larval body temperatures and observing larvae in the field to estimate the proportion of time spent in sunlight at various ambient temperatures. Another possible approach would be to use data from field populations in the construction of the model, perhaps by means of such a program as DEVAR, so that patterns of behavior are built into the model. However, a single relationship may not account for behavioral thermoregulation if there are large seasonal differences in behavior.

3. Differences Under Fluctuating Temperatures

Discrepancies are common between development times predicted from constant development-temperature relationships and those measured under "natural" fluctuating temperatures around the same mean. Ratte (1985) summarizes these studies and in a simulation experiment using a nonlinear rate-temperature relationship, shows that rates under fluctuating temperatures will be slower at higher, and faster at lower temperatures than the mean constant temperature. Pradhan (1945) provides a similar graphical explanation of how degree-day models frequently underestimate development rates under fluctuating temperatures because of rate summing with nonlinear functions.

Foley (1981) found faster development of *H. armigera* pupae at fluctuating than at constant 24°C. Ratte (1985) shows that this discrepancy may be the result of rate summing with a nonlinear function that accounts for 58% of the reported experimental discrepancies. The remaining 42% may result simply from the use of an inappropriate model. Even when curvilinear relationships are used, some deviation may still be apparent (Tanigoshi, Browne, Hoyt, and Lagier, 1976; Gregg, 1982; Allsopp, 1986). Other physiological and behavioral responses to thermoperiods *per se* may be implicated (Beck, 1983). If pronounced these will preclude the use of constant-temperature data to predict development under fluctuating temperatures, whether or not a nonlinear model is used. Gregg (1982) and Allsopp (1986) found discrepancies that were not explained by constant-temperature nonlinear models for the Australian plague locust and false wireworms, respectively. The existence of such difficulties with *Heliothis* spp. has never been resolved satisfactorily, although the data of Eubank, Atmar, and Ellington (1973) suggest that there are no major differences for the eggs of *H. zea*. To resolve the question would require either the comparison of model predictions with development times under controlled fluctuating temperatures, for example, Gregg (1982), or the fitting of nonlinear models to result from constant and fluctuating regimes using a numerical optimization algorithm such as DEVAR. If intrinsic thermoperiod effects occur, the model derived from fluctuating-temperature data will be quite different to one derived from constant-temperature data (Dallwitz, 1984; Allsopp, 1986).

4. Host Plant Variation

Another large source of developmental variation in the field is variation in host plants, either in defensive chemistry or nutritive value. Most work with *Heliothis* spp. has utilized artificial diet. Development tends to be faster on artificial diet than on most plants (Pretorius, 1976; Butler, 1976). In addition, larval development of *H. armigera* can be extended by an additional 1 to 2 instars over the normal 5 to 6. Twine (1978) suggested a role for diet in this effect. Kay, Noble, and Twine (1979) showed that the addition of even low concentrations of the terpenoid pigment gossypol greatly extended larval development times in *H. armigera* and *H. punctigera*. Heard (1985) found that larval development is slowed if the insects are reared on artificial diet to which extracts of damaged plant parts have been added.

Laboratory studies of differences in development of *H. armigera* (Coaker, 1959; Pretorious, 1976), *H. zea* (Butler, 1976), and *H. virescens* (Nadgauda and Pitre, 1983) on different plants and different cotton plant parts (buds, bolls, and leaves) have failed to quantify the biochemical basis of the response and thus generalize the results. Even if such variables as nitrogen content and defensive chemistry were measured and entered into the development functions as for salvinia moth (Taylor, 1984), active food choice may still make generalization to field conditions difficult. Brier and Rogers (1981) found that small larvae of *H. armigera* fed on small leaves of soybean before shifting onto pods that seem to be tougher but higher in protein. Similarly, Wilson and Waite (1982) found that older larvae fed selectively on older cotton fruit. As with behavioral temperature compensation, the effect of behavioral flexibility in feeding can be determined only by measurement under a variety of field conditions.

5. Intraspecific Variability

As has been shown for many insects, *H. zea* (Sharpe, Schoolfield, and Butler, 1981), the temperature-development relationship may show considerable between and within population variation. After several generations of laboratory rearing, *H. armigera* can evolve quite different characteristic temperature-development relationships (Twine, 1978). Similarly, Brier and Rogers (1981) found that field populations had lower developmental variability than did laboratory populations. This source of error limits generalization to field populations of data collected using laboratory strains or individuals from one population. There are also differences in developmental rates between the sexes (Reed, 1965; Foley, 1981). These differences are relatively small and are significant for population dynamics only in the unlikely circumstances of highly skewed sex ratios.

6. Summary of Sources of Variation

Of the sources of error in prediction, viz. microclimate, thermal heterogeneity, thermoperiod physiological effects, behavioral thermoregulation, host plant effects, and intraspecific variation, the first two are essentially technical problems. They are solved readily by microclimate measurement and fitting nonlinear models for development rates under fluctuating temperatures. Behavioral thermoregulation and host plant effects are highly variable and depend on both plant phenology and weather, and intraspecific variation is always present. Improvement in the local predictability of population phenology, therefore, requires measurement of life histories of local populations feeding on local plants.

V. Temperature-Related Studies with Australian *Heliothis*

Kirkpatrick (1962) made the first records of the time spent by *H. armigera* and *H. punctigera* as eggs, larvae, and pupae. All stages were reared inside an insectary and development times related to mean monthly temperatures. In midwinter (May–July) with temperatures averaging 16 to 18°C, both species took ca. 73 days to develop from egg to adult, assuming no pupal diapause occurs. In summer (December–January) with temperatures ca. 28°C, both species took ca. 34 days. These data give only an outline of the species development and it is unclear what diet was used and in some months records are missing.

Cullen (1969) studied the effects of a range of constant temperatures (9.5–40.0°C) on the development of all stages of *H. punctigera* reared on chopped French beans. Optimal development occurred at ca. 35°C for all stages and he estimated (by eye) a developmental threshold of 10°C. Females developed at a faster rate (30.74 days) than did males (32.46 days) at 19.4°C.

Twine (1978) found no sex difference in *H. armigera* development when reared on artificial diet. He determined the effects of constant temperatures ranging from 13.1 to 38.4°C on the development of larvae and pupae and estimated, for the combined larval and pupal stages, a developmental zero of 11°C and a thermal constant of 475 degree-days. Maximum rate of development occurred at 33.9°C.

Wilson et al. (1979) studied the effects of constant temperatures on field-collected autumn pupae of *H. armigera* that were collected as late instar larvae. For nondiapause and postdiapause development they calculated a developmental threshold of 12°C from a curve fitted by eye to the inverse of the median time for the pupal period.

In another study of constant temperatures on development of *H. armigera* pupae, Foley (1981) reported a developmental threshold of 14.8 ± 0.96°C for nondiapausing pupae and a thermal constant of 160 degree-days, considerably less than the 200 degree-days determined by Wilson et al. (1979) and 211 degree-days of Twine (1978). These differences are probably mostly due to the choice of different threshold values, that is, to different interpretations of the development rate curve rather than to intrinsically different experimental results.

Kay (1981) reared eggs of *H. armigera* under constant temperatures (range 8.0 to 39.4°C) and found a developmental threshold of 11.7°C and a thermal constant of 43.3 degree-days. Maximum rate occurred at 33.9°C as in larvae and pupae (Twine, 1978).

Room (1983) used the data of Twine, Kay, and Cullen to derive expressions for development of immatures and adult females to 50% oviposition against temperature. He used Pradhan's (1946) equation with an optimum temperature of 35°C for all stages. The same expressions were used for both *H. armigera* and *H. punctigera* as he considered that their responses to temperatures were similar. Using screen temperatures experienced by eggs and larvae on cotton and pupae in soil, he calculated the times for development through each stage. The predicted generation times agreed with field observations of peak egg counts and peak light trap catches in the Narrabri area (Wardaugh et al., 1980).

Chapter 9

Methods for Studying Diapause

D.A.H. Murray and A.G.L. Wilson

I. Introduction

The term *diapause* has been used variously to cover the delays in development of organisms that allow them to survive unfavorable conditions. It is important to make the distinction between true diapause where development is arrested spontaneously and does not respond immediately to amelioration of the external environment, and quiescence, where development is temporarily inhibited by an unfavorable environment and may be resumed as soon as the hinderance is removed (Andrewartha, 1952). Both these conditions occur with *Heliothis* spp. and have important different effects on the time of moth emergence. The abundance of *Heliothis* spp. in a particular year can be influenced greatly by timing of entry and exit from diapause.

II. Why Study Diapause?

Diapause is the most important mechanism for overwinter survival of *H. armigera* and possibly *H. punctigera* in subtropical Australia. Knowledge of the factors that induce and terminate diapause will assist interpretation of the temporal patterns of diapause incidence and aid the prediction of post-diapause populations. These studies are essential for the construction of population models.

The questions such studies need to answer include: At any point in time, What proportion of the population will enter diapause? and When will the moths emerge?

Field investigation of the timing, location, and abundance of diapausing pupae is also of importance in studies of the effect of cultivation practices on population survival (Roach, 1981). Such studies may have the aim of suppressing overwintering populations and in particular, the gene pool for insecticide resistance in *H. armigera*.

A further line of research may be to determine the extent of parasitism, predation, and disease on pupal survival as part of population dynamics and control studies.

III. Recognition of Diapause

Diapause studies in *Heliothis* spp. have involved both laboratory and field investigations. Fundamental to these studies is the recognition of the diapause condition in pupae.

Two criteria are used to determine diapause in *Heliothis* spp.: (1) the visible characters of arrested development and; (2) the duration of the pupal stage. Shumakov and Yakhimovich (1955) described the presence of eyespots in the postgenal region of newly moulted pupae of *H. armigera*, and their disappearance during subsequent development. The same phenomenon occurs in other *Heliothis* spp. (Phillips and Newsom, 1966; Cullen and Browning, 1978).

Immediately following larval–pupal ecdysis the eyespots are distinct, occurring in a straight line across the postgenal area (Fig. 9.1A). During development of the pharate adult their prominence and position change (Fig. 9.1). Five stages can be recognized, whose durations in days in uninterrupted development at 28°C are (A) 2–5; (B) 1; (C) 2–3; (D) 1; and (E) 2–3. Diapause when present always occurs in Stage A. Once Stage B occurs, development proceeds without interruption until the adult emerges. For any temperature, the duration of Stage B to eclosion varies little, most of the variability in the time between larval–pupal ecdysis and emergence of the adult being due to the variation in the time spent in Stage A. The retention of pigmented eyespots (stemmata) beyond the normal period is a characteristic of diapausing pupae. Similarly, the fat body of newly formed pupae is composed of firm rounded lobes and remains unchanged in this condition throughout diapause (Pearson, 1958). In nondiapausing pupae, histolysis starts on the second day and the fat body disintegrates into a mass of free granules that later reform into the adult fat body.

Diapause manifests itself by a prolonged period between pupation and adult emergence. Under constant temperature conditions, diapause is considered to occur if emergence takes longer than a predetermined period. If emergence took longer than 20 days at 28°C or 50 days at 19°C, *H. punctigera* were considered to have been in diapause (Cullen and Browning, 1978). As the criterion for diapause in *H. armigera*, Hackett and Gatehouse

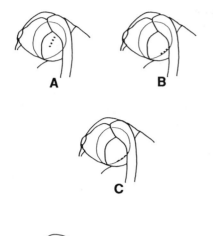

Figure 9.1. Stages in movement and disappearance of the pigmented eyespots during development of *H. punctigera* pupae. Five stages (A–E) can be recognized, whose durations in days during development at 28°C without diapause are A, 2–5; B, 1; C, 2–3; D, 1; E, 2–3. (Cullen and Browning, 1978). (Redrawn with permission from Pergamon Press.)

(1982a) used an extension of the pupal period longer than 1.5 times that of the minimum pupal period they recorded for nondiapause pupae. Duration of development is a satisfactory criterion of diapause for individuals reared under identical (constant or fluctuating) temperatures, but cannot be used for individuals reared at different times at variable fluctuating temperatures. In this situation, it is preferable to determine the thermal summation for pupal development of each individual, using daily maximum and minimum (or hourly) temperatures (see Chapter 8). Those individuals with thermal summations greater than say 1.5 times that of the minimum value recorded for nondiapause pupae can be considered to have been in diapause.

1. Problems with Diapause Recognition

Most individuals have a full complement of four eyespots in each postgenal region. In some pupae, the eyespots can be very faint and not discernible. In these cases, it is useful to check on the condition of the fat body in order to determine the diapause condition.

The duration of pupal development in days or thermal units has been used to distinguish nondiapausing and overwintering *H. punctigera* and *H. armigera*. *H. punctigera* also enters a spring diapause (Cullen and Browning, 1978) from which the timing of emergence is extremely variable.

Due to this variability a number of individuals become borderline diapause cases on the duration criterion and consequently their diapause condition is not definite.

IV. Laboratory Studies

Laboratory studies have examined factors affecting diapause induction and termination. True diapause appears to be determined by photoperiod and temperature. A third factor, diet, has a minor role. A very narrow range of photoperiod (between 11.5 and 12.5 h day length at 19°C) induces diapause in *H. punctigera* (Cullen and Browning, 1978). In contrast, the diapause-inducing range of photoperiod for *H. armigera* in Africa (Roome, 1979) and Russia (Komarova, 1959) is wider than that found in Australia for *H. punctigera*. As well as the normal incidence of diapause during late summer-autumn when photoperiod is decreasing, *H. punctigera* enters diapause during the spring when photoperiod is increasing (Cullen and Browning, 1978).

A decrease in photoperiod between that experienced by the egg and that experienced by the larvae increases diapause induction (Cullen and Browning, 1978). In *H. punctigera* and *H. armigera* high temperatures (28°C and above) override diapause even at optimum photoperiods, whereas low temperatures (18–22°C) increase diapause induction.

High temperatures are known to induce summer diapause in *H. virescens* (F.) in Arizona (Butler, Wilson, and Henneberry, 1985) and a similar phenomenon has been observed for *H. punctigera* in the Namoi Valley and *H. armigera* in the Ord of Australia (A.G.L. Wilson, unpublished). Factors inducing quiescence have not been investigated.

After a period of delayed development, diapause in *H. armigera* is terminated at a theoretical threshold temperature of 16.36°C (Foley, 1981). Once terminated, postdiapause development proceeds at a rate similar to that for nondiapause pupae (Wilson, Lewis, and Cunningham, 1979). The calculated temperature threshold for diapause termination in *H. zea* and *H. virescens* are 16.41 and 15.65°C, respectively (Lopez, 1986). Accurate determination of these thresholds and the distribution of emergence from diapause are useful for the construction of emergence models (Logan, Stinner, Rabb and Bacheler, 1979; Cunningham, Lewis and Wilson, 1981).

1. Problems with Laboratory Studies

Most laboratory investigations into diapause induction use controlled environment cabinets, in which it is not possible to simulate field conditions of reducing or increasing photoperiod and fluctuating temperatures accurately. Studies using environment cabinets are most useful in examining such main effects as photoperiod and temperature, but caution should be exercized when extrapolating from laboratory data derived in this way. Despite their shortcomings, environment cabinets probably will continue to play an important part in our endeavors to understand diapause.

Parental stock and the conditions under which they are held will influence the outcome of diapause induction experiments, so this aspect deserves careful consideration. Always maintain parental stock at a photoperiod longer than that which is under investigation for egg and larval development. High levels of diapause induction may be very difficult to obtain at constant photoperiods (Wellso and Adkisson, 1966).

V. Field Studies

Field studies into diapause have used several methods. The seasonal incidence of diapause has been studied by collecting mature larvae from crop hosts, and rearing them through in an open-air insectary on artificial diet or host plant material, or by maintaining cultures in a similar way. The possible influence of diet must be borne in mind. As it is necessary to examine pupae to determine their diapause condition, they can be held individually in containers with a small quantity of vermiculite (Kay, 1982a). Alternatively, the pupae are placed in clear glass tubes plugged with cotton wool so that they can be examined without further direct handling (Wilson et al., 1979). It may be preferable to hold pupae in tubes buried in soil trays or pots so that temperatures they experience are more comparable to field temperatures.

Field emergence of adults can be investigated using tubes as artificial burrows (McCann, Lopez and Witz, 1989). Diapausing pupae produced in the laboratory or collected from the field, are placed in tubes at specific soil depths in the field. Moth emergence from the tube into a small collecting bottle is recorded daily and accurate temperature records obtained.

Alternatively the mature larvae are released into field cages where environmental conditions are more natural. Pupae are excavated during the season to determine diapause condition or the time of moth emergence can be used as the criterion for diapause (Wilson et al., 1979).

Pyramid cages (1 m²) (Shiller, 1946) have been used very extensively to investigate diapause in the field. Mature larvae—either field-collected or reared in an open-air insectary—are released onto the soil surface and allowed to enter the soil and pupate.

Small cages—one larva per cage—are used to obtain more detailed information on an individual basis. One successfully used small cage is an open-ended cylinder of wire gauze (fly screen mesh), 7.5 cm in diameter and 15.0 cm long. This cage is placed in soil to a depth of about 10 cm and the inner soil core is retained intact. The cage is capped with a gauze lid to prevent the escape of larvae.

The development of pupae is often atypical in cabinets or cages, where the parameters affecting development may not be reproduced fully or known. Excavation of natural pupal populations in crop residues is an alternative method used to investigate diapause (Wilson et al., 1979). The surface soil is scraped carefully with a trowel to reveal emergence tunnels and the pupae then are excavated carefully (see Chapter 5). Favored late summer-autumn host crops such as pigeon pea usually will provide large overwintering populations.

Sequential excavations of field or cage populations will provide data on survival (Slosser, Phillips, Herzog and Reynods, 1975; Eger, Sterling, and Hartstack, 1983). Alternatively, field cages are placed over field plots to monitor moth emergence. Specific survival experiments, for example, to examine the effect of cultivation, may rely on sampling pupal populations before and after treatment (Wilson, 1983) or the placement of field cages over treated and untreated areas to monitor moth emergence.

1. Problems with Field Studies

A number of problems arise if mature larvae are released into field cages. It is important that the larvae are at the correct stage of development for release. If larvae are too mature they will not tunnel into the soil. Consequently, they will pupate on or near the soil surface. If larvae are not mature, they continue to feed for a day or two or wander around the cage in search of food before eventually dying. When larvae continue to feed in the diet cups they experience very high temperatures during the day time. The diet will dry out and the larva is forced to abandon feeding. It is best to release larvae in the late afternoon so that they have about 12 h during the cooler part of the day in which to complete feeding and tunnel into the soil. Larvae are suitable for field release within 24 h of onset of tunnelling into the diet to begin construction of the pupal cell.

Cannibalism occurs in the pyramid cages if too many larvae are released. Fifty to 100 larvae /m^2 of cage have been released successfully. Mice and ants are sometimes troublesome, preying on prepupae and pupae inside cages. It is difficult to exclude mice if the soil is dry as they enter the cages via cracks in the ground. Ants are active foragers and are most troublesome in the individual wire gauze cages. Ant nests can be treated with insecticide or controlled by poison baits.

Field-collected larvae are sometimes used for cage release but parasitoids reduce sample sizes, a problem if numbers are critical., On the other hand, field-collected larvae are useful if overwintering of parasitoids is the topic of investigation.

With the pyramid cages, such data as date of entry into soil, depth of pupal cell, and date of moth emergence are not obtained on an individual basis, whereas the small cages will provide these records. Pupae in cages are excavated successfully if soil moisture is suitable—soil too wet or too dry makes excavation difficult. Excavation is useful to determine numbers of unemerged pupae, or dead moths in emergence tunnels, and thus, obtain an accurate assessment of survival.

Although both cage types are suitable for within crop studies, it is necessary to remove most of the plant material within the pyramid cages to facilitate their placement. Removing plants may change microhabitat temperature conditions. The small cages can be placed wherever desired with no need to remove vegetation. This aspect is not particularly important in diapause studies, because for the most part, vegetation is absent during the winter–spring study period, but it may be necessary to remove plants in the autumn when larvae are released.

Natural populations of diapausing pupae are frequently at low densities ($< 1/m^2$) so field cages need to be relatively large to provide sufficient numbers of emerging moths. Large cages have the inherent problems of wind resistance and susceptibility to hail damage, conditions often encountered in some areas during the spring–early summer months.

All cages modify the surrounding environment so it is important to determine what modifications are imposed. This is especially important because soil temperature will influence termination of diapause and subsequent pupal development.

Temperature probes and data-logging equipment are ideal for this purpose. Because of the soil thermocline and variable depth of *Heliothis* pupae, it is desirable to obtain soil temperatures from at least two depths—say, 2.5 and 5.0 cm. These should be compared with soil temperatures outside cages and air temperatures. These parameters will assist data interpretation and the understanding of the diapause process greatly.

Chapter 10

Methods for Studying Adult Movement in *Heliothis*

V.A. Drake

I. Introduction

Movement is a key factor in the population processes of any mobile insect (Dingle, 1972; Horn, 1983; Solbreck, 1985; Stinner, Saks, and Dohse, 1986). Movements into and out of a population will affect its density, its age structure, and its genetic composition (Hughes, 1979). Information about capacity for movement is essential for the development of strategies for managing an insect pest.

In *Heliothis*, for which there is evidence of migration for a number of species (Farrow and Daly, 1987; Fitt, 1989), information about movement is required specifically for the development of

1. population forecasting. Immigration, whether from neighboring fields and field margins, or from distant cropping regions, pasturelands, and native vegetation may predominate over local production; and locally generated adults may emigrate before ovipositing (Topper, 1987a). Recognition of invasion and prediction of emigration is central to the short-term (approximately single-generation) forecasting methods that are used in the control of highly mobile insects (Betts, 1976). Forecasting beyond a single generation will require a model of the species' population dynamics that incorporates both movement and spatial heterogeneity (Wellington, Cameron, Thompson, Vertinsky, and Landsberg, 1975; Taylor and Taylor, 1977; Clark, 1979; Royama, 1980; Stinner et al., 1986),
2. conventional control. Incorporation of information on immigration and emigration processes (especially on their scale and likelihood of occur-

rence) into the design of conventional chemical control strategies could lead to improved operational efficiency (Haggis, 1982),

3. area-wide management. Movements into and out of the managed area will determine the effectiveness of an area-wide (multiple-generation) control scheme (Knipling and Stadelbacher, 1983; Mueller et al., 1984). The scale of movements will determine the size of a buffer zone required around the area to be protected, and thus the minimum size of the managed area (Rabb, 1979),

4. alternative management methods. Alternatives that act on the adult stage, such as sterile insect release (Proshold, Laster, Martin, and King, 1982), mating disruption by synthetic pheromones (Sparks et al., 1982), chemical repellents (Saxena and Rembold, 1985, see Chapter 11), trap crops (Sloan, 1938; Topper, 1987a), and insecticidal control of adults (Joyce, 1982a) all depend to some extent on the frequency and scale of adult movements, and

5. management of insecticide resistance. Movement alters the spatial distribution of the genes conferring resistance and, therefore, affects its evolution, maintenance, and spread (Comins, 1977; Daly and Fitt, 1990).

II. Types of Movement

Movements of *Heliothis* adults occur on a range of scales and have a variety of ecological consequences (Farrow and Daly, 1987). In the present context, a classification by scale appears most appropriate.

1. Within-field movements (characteristic distance of order 100 m) will occur due to the "trivial" (Southwood, 1962) flights associated with such appetitive behavior as feeding, finding shelter, and reproduction. Such flights are likely to be made within or immediately above the crop canopy, at altitudes where the moth's airspeed exceeds the speed of the wind, that is, within the species' "flight boundary layer" (Taylor, 1974). In a broadacre agroecosystem, most movements on this scale will occur within the basic management unit (the field). In smaller-scale systems, such as a market garden or an intercropped field, even these short-distance movements may carry moths between hosts of different types.

2. Between-field movements (characteristic distance of order 1 km) allow colonization of field crops by moths generated nearby on other hosts. For moths invading cotton, neighboring groundnut (Joyce, 1982a; Topper, 1987a,b) and wheat (Wardhaugh et al., 1980) crops, and weeds growing along field margins and rights-of-way (Stadelbacher, 1981), have all been identified as sources of infestation. Movements on this scale require only about 5 to 10 min of flight and are made within or just above the moths' flight boundary layer, at altitudes below 10 m (Joyce, 1982b). Such flights may be appetitive as well as dispersive, as when moths emerging from a senescing host invade an adjacent crop that provides a source of nectar.

3. Within-region movements (characteristic distance of order 20 km) lead to

mixing of populations within areas in which climatic conditions, and therefore, crop and pest phenologies, are approximately synchronous. Movements on this scale would also result in exchange of populations between neighboring asynchronous ecosystems, for example, between an irrigated area and nearby dryland crops or unimproved pastures. Such movements could be achieved either by extended flights, probably wind-assisted, at heights of 10 m or less (Brown, 1970), or by brief downwind flights in the faster-moving air at higher altitudes.

4. Between-region movements (characteristic distance of order 200 km) enable exchanges between populations inhabiting widely separated areas that may have different climates and in which the phenologies of crops, wild hosts, pests, and natural enemies may be asynchronous. Such movements can lead to outbreaks occurring in areas where the species cannot overwinter (Blanchard, 1942; Fitt, 1989) and to early spring infestations arising well before the emergence of local populations (Hartstack et al., 1982; Fitt and Daly, 1990). Movements over these distances in midseason could cause the failure of an area-wide *Heliothis* management strategy by reinfesting regions in which control measures have been used to suppress early-season populations (Mueller et al., 1984). A lack of geographic variation in gene frequencies provides indirect evidence that population mixing occurs on this geographic scale (Daly and Gregg, 1985; Gunning and Easton, 1989, see Chapter 13). Movements between regions can be achieved by nighttime flights at altitudes of a few hundred meters (Callahan et al., 1972). Distances of 100 to 300 km will be covered in a single flight of a few hours duration (Drake and Farrow, 1985) in environmental conditions that may be ideal for long-distance migration (Drake, 1984). The capacity of *Heliothis* for such flights is demonstrated by its ability to make sea crossings of a few hundred kilometers (Sparks et al., 1975; Drake, Helm, Readshaw, and Reid, 1981; Farrow, 1984).

5. Extra-limital movements (characteristic distance of order 1000 km) carry small numbers of moths well outside the species' breeding range (Fox, 1975; Holloway, 1977; Pedgley, 1985). Such movements are of no ecological significance as the immigrants soon perish.

III. Methods for Studying Movement

Five general methods have been used for the detection and measurement of the various types of movement described above. Each of these methods has a number of variations, many of which have been tried on *Heliothis* or other noctuids.

1. Mark-and-Capture

Mark-and-capture studies provide firm information about movements of individual insects between known localities. They can be used on any scale, but capture rates at the destination are very low for long-distance movements and large numbers of insects must, therefore, be marked in the source areas (see Chapter 5). The moths can be marked externally, for example, with a felt-tip pen, or internally with dyes, radioisotopes, or rare elements (Raulston, 1979).

For studies of natural populations, marking without capture, for example, by introducing a radioactive (Snow et al., 1969) or rare-element (Van Steenwyk, Ballmer, Page, Ganje, and Reynolds, 1978) tracer or a dye (Bell, 1988) into the larval food plant or by providing dye-marked baits for adults (Rose, Page, Dewhurst, Riley, Reynolds, Pedgley, and Tucker, 1985) has the advantage of enabling large numbers of moths to be marked with minimal disturbance. If the alternative capture-recapture method is used, it is relatively easy to employ a series of marks for different release points and dates. This method may be best suited to the study of short-range movements where recapture rates are relatively high so that fewer moths need to be marked; however, it can also be used to study long-distance movements (Raulston, Wolf, Lingren, and Sparks, 1982). Such studies may be able to provide an estimate of the population turnover at the trapping site (Farrow and Daly, 1987).

Mark-and-capture experiments using laboratory-reared moths (Hendricks et al., 1973b; Haile et al., 1975) can provide basic information about the species' capacity for movement. Release of genetically modified moths carrying a hereditary mark enables the detection of movement over a number of generations (Bartlett and Raulston, 1982). These methods are of value for investigating the dispersal of insects released in autocidal control programs, but only limited inferences can be drawn from them about the behavior of naturally occurring populations.

2. Capture of Naturally Marked Specimens

A variant of mark-and-capture is to infer the origin of captured specimens from naturally occurring marks. This provides information about immigration of individual insects into a known locality from a general source area. The method has the practical advantage that the task of artificially marking a sufficient population in the source area is avoided, but considerable research may be required to establish the reliability of the naturally occurring mark and the range of localities in which it originates. Natural marks that might enable *Heliothis* movements to be detected include disease (Wellington, 1962), parasites (Treat, 1979; Hughes and Nicholas, 1974), pollen (Hendrix, Mueller, Phillips, and Davis, 1987), and elemental composition (Bowden, Brown, and Stride, 1979; Zhulidov, Poltavskii, and Emets, 1982); the latter two may also provide information about movements between different host species (Mikkola, 1971; Sherlock, Bowden, and Digby, 1985).

Even characters like size, age, or degree of wear can sometimes be used to detect immigration (Hughes and Nicholas, 1974; Holloway, 1977; Bowden et al., 1979). Genetic markers such as insecticide resistance (Wolfenbarger, Lukefahr, and Graham, 1973; Gunning and Easton, 1989) and allozymes (Sell, Whitt, Metcalf, and Lee, 1974; Pashley and Bush, 1979; Daly and Gregg, 1985; Daly, 1989) can indicate whether populations are isolated from each other or whether there is mixing in one or both directions, but they cannot usually provide information about individual movements.

3. Observation at Times or Places Where the Moths Could Not Have Been Produced Locally

Capture or observation at a locality surrounded by habitat that is completely unsuitable for breeding provides information about the distance the species is capable of moving. Most satisfactory for this purpose is observation on marine platforms (Sparks et al., 1975) or ships at sea (Hsia, Tsai, and Ten, 1963; Wolf, Sparks, Pair, Westbrook, and Truesdale, 1986), as there is then no possibility of local production. Observation on offshore islands (Holloway, 1977; Farrow, 1984), and even on neighboring landmasses (Fox, 1975), can be interpreted in the same way if it can be established that the species concerned is not endemic. Insects that have perished before reaching such localities can sometimes be detected by searching the tideline of beaches for corpses (Farrow, 1975; Drake and Farrow, 1985).

Observation in habitat suitable for breeding can provide evidence of immigration in some circumstances. For example, observation in areas where the species does not overwinter (Blanchard, 1942; Pedgley, 1985), or early in the season when the local population has not yet emerged (Fitt and Daly, 1990; Hartstack et al., 1982), are clear indications of immigration. If preceding larval populations have been monitored or controlled, immigration can also be inferred from the appearance of adults in numbers greater than could be produced by local emergence (Greenbank, 1957; Wilson, 1983). Monitoring of adult populations alone can give an indication of immigration by identifying sudden increases in numbers that cannot be accounted for by local factors (Morton et al., 1981). This method, which requires a very careful analysis if the results are to be reliable, may also detect emigration.

An extension of these methods is to use the "synoptic survey" technique (Joyce, 1982b), that is, monitoring over a wide area, to identify regional patterns of immigration (Brown, Betts, and Rainey, 1969; Haggis, 1982). Areas in which local production is insufficient to account for increases in population between generations can be identified as having received migrants. Areas producing adults at the time of migration are candidate source regions for these migrants. It may even be possible, with appropriately timed observations, to detect emigration from a decrease in an adult population. The dates at which populations are in the adult stage, that is, at which migration can occur, can be estimated from observations of immature populations by using a development model (Oku, 1983; Smith and McDonald, 1986, Farrow and McDonald, 1987, see Chapter 8), whereas a regional population model (Stinner et al., 1986, Chapter 14) may be the most effective tool for integrating and interpreting synoptic survey data. Monitoring surveys can similarly be used to detect short-range movements between hosts with different phenologies, that is, between-field movements (Wardhaugh et al., 1980; Joyce, 1982a; Topper, 1987a).

4. Capture During Movement

Capture of moths in flight provides firm information that movement was occurring at a particular locality, but almost no information about the source or destination. For studies of trivial movements, capture methods that rely on appetitive behavior, for example pheromone traps, may have some application, but for the nonappetitive migratory movements, capture should be by interception during flight. Suction and rotary traps (Southwood, 1978) and nets flown on remotely piloted vehicles (Tedders and Gottwald, 1986) are of little value for *Heliothis* research because catch rates would be too low. Interception at low altitudes can be achieved by large, stationary flight traps (Walker, 1985) oriented with their aperture facing the direction of migration, or by a large net mounted on a vehicle (Downing and Frost, 1972). Giant versions of both these types of trap have been developed by the Division of Entomology, CSIRO, in Canberra, Australia, for sampling the low density populations that are typical of *Heliothis* (Figs. 10.1, 10.2). Interception of moths being carried downwind at altitudes of 100 to 200 m can be accomplished with a kite-borne net (Farrow and Dowse, 1984). Interception at higher altitudes requires a net mounted on an aircraft (Glick, 1965, Reling and Taylor, 1984). An important advantage of all these interception methods is that they enable insect density to be estimated by an absolute method (Farrow and Dowse, 1984). A disadvantage is that catch-rates for *Heliothis* moths are usually very low (Glick, 1965; Drake and Farrow, 1985).

One alternative to interception is to use light-traps mounted well above ground level and shielded so as to be visible only from above (Taylor, Padgham, and Perfect, 1982). A series of such traps mounted on a very tall tower caught several thousand *H. zea* adults over two seasons of operation, giving an indication of the distribution of flying heights for this species (Callahan et al., 1972). Variations in trap efficiency due to environmental factors (Morton et al., 1981; see Chapter 4), and uncertainty about the mechanism of long-distance attraction of moths to light (Hsiao, 1972), make light trap catches relatively difficult to interpret.

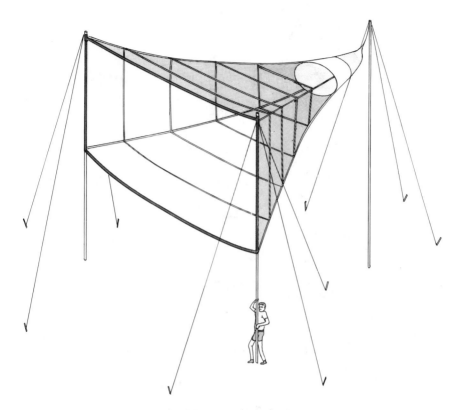

Figure 10.1. "Goal-net trap" developed by CSIRO Division of Entomology for sampling moths flying downwind above crops at heights of 3–6 m.

5. Observation of Movement in Progress

A second method of obtaining information about movement at a particular locality is by observation, either visually, usually with the aid of an optical or opto-electronic device for enhancing vision, or by electronic remote-sensing techniques. In comparison with capture, this method has the advantage that it is usually possible to study a reasonable number of individuals, and that information on the moths' direction and height of flight can be obtained. Its most important disadvantage is that *Heliothis* moths flying at night are difficult to identify by observation alone. The use of field cages stocked with laboratory-reared or locally captured populations overcomes this identification problem and enables the study of some aspects of flight activity, for example, timing (Lingren, Greene, Davis, Baumhover, and Henneberry, 1977; Oku, 1980), in near-natural environmental conditions. Initial dispersal and emigration flights can be observed without cages by placing roosting or newly emerged moths at known locations before dusk

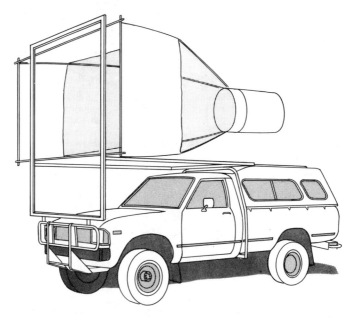

Figure 10.2. Giant vehicle-mounted net developed by CSIRO Division of Entomology for sampling moths flying between crops at heights of 2–4 m.

(K. Hirai, personal communication). However, observations of most aspects of *Heliothis* movement will need to be done on wild populations that have not been subject to any experimental manipulation.

Visual observation can be used to study emigration ("take-off") flights at dusk by using binoculars to scan the western sky, provided this is clear of cloud (Drake and Farrow, 1983; Riley, Reynolds, and Farrow, 1987). Take-off flights can be seen throughout the night with the aid of a night vision telescope (Greenbank, Schaefer, and Rainey, 1980). Nocturnal flight just above the crop canopy can be studied with a torch (Topper, 1987b) or with nightvision devices and red-light illumination (Lingren, 1978b). The latter technique should eliminate disturbance of the moths' behavior. A narrow horizontal-illumination beam and marker-poles can be used to delineate an observation volume for quantitative studies (R.A. Farrow and G.P. Fitt, personal communication). A vertical illumination beam enables the direction and intensity of movement to be quantified at heights up to 10 m (Brown, 1970), and probably to at least 30 m if nightvision devices are used (Lingren et al., 1978b). Flight trajectories could be determined by simultaneously tracking an individual with a pair of sighting devices (Zalucki, Kitching, Abel, and Pearson, 1980). Moths flying at higher altitude can be observed through a telescope as they pass across the face of the moon

(Pruess and Pruess, 1971). Occasionally visual observations have detected descent flights (Greenbank et al., 1980).

Photographic or video techniques, preferably combined with infrared illumination, can be used to record movements for subsequent analysis. Image-intensified cine (Murlis and Bettany, 1977; Riley, Smith, and Bettany, 1990) or infrared-sensitive (Howell and Granovsky, 1982; Schaefer and Bent, 1984) video systems can provide a sequence of images from which trajectories can be calculated. Modern image-intensified video systems require only low-intensity infrared illumination and are well suited to field use (J. Murlis, personal communication). Multiple-exposure photographs that can provide estimates of speed, track, and orientation of low-flying insects (Sayer, 1956), could be obtained at night by firing a rapid sequence of flashes while the camera shutter remains open.

Moths flying through an infrared beam can be detected electronically without forming an image. A device developed by Riley, Reynolds, and Farmery (1981) was capable of detecting moths at heights up to about 30 m. It could also provide measurements of the wing-beat frequencies of the targets it detected, thus giving some indication of their identity.

These optical methods enable flight within a few tens of meters of the ground to be studied, whereas observations at higher altitudes can be made with radar (Schaefer, 1976; Riley, 1989). Special-purpose "entomological" radars are capable of detecting *Heliothis*-sized moths at heights from a few meters above the vegetation canopy to over 2 km, and of measuring migration parameters quantitatively to an altitude of 1 km (Riley, Reynolds, and Farmery, 1983; Drake and Farrow, 1985). Radars of the scanning, pencil-beam type provide measurements of insect density, track, and speed as a function of height, and migration fluxes, the total migration rate, and the total number of insects that have passed overhead, can be estimated from these (Schaefer, 1976; Drake, 1982a; Chen, Bao, Drake, Farrow, Wang, Sun, and Zhai, 1989). These radars are also capable of detecting emigration flights of moths taking off in plumes from an infested crop (Schaefer, 1976; Lingren and Wolf, 1982) or a localized outbreak site (Riley et al., 1983). A great advantage of this radar configuration is its ability to scan a volume of order 10^7–10^8 m^3 in a few seconds, so that low-density migrations that are undetectable by other methods can be studied. Vertical-beam entomological radars (Riley and Reynolds, 1979; Bent, 1984) have a smaller sensitive volume, but are capable of unmanned operation, with data-acquisition under computer control, and may be suited for long-term monitoring studies. Airborne entomological radars (Schaefer, 1979) can measure migration density as a function of height over a transect hundreds of kilometers long.

Radar can provide information about the identity of the targets it is detecting through measurements of wing-beat frequencies (Schaefer, 1976; Drake and Farrow, 1983; Riley et al., 1983) and, in the case of vertical-beam radars, size (Bent, 1984) and body shape (Riley and Reynolds, 1979). However, identifications from these measurements are likely to be limited to distinguishing moths of *Heliothis* size from other insects. The inability to provide specific identifications is a serious shortcoming of all of these remote-sensing techniques.

IV. Supporting Evidence for Movement

In addition to the above methods of studying movement directly, there are a number of observations and experiments that provide supporting evidence for movement. Data from some of these ancillary procedures are also of great value when interpreting observations of movement obtained by more direct means.

1. Behavioral and Physiological Studies

Laboratory studies of captive moths give an indication of the species' capacity for movement by determining the duration of its flights; they also enable the factors affecting flight capacity to be identified (Hackett and Gatehouse, 1982b). Laboratory or fieldcage observations of the physiological development of the young adult, and of the timing of reproductive and feeding activity as well as flight, can be integrated into a descriptive model of adult behavior (Topper, 1987b). Physiological examination of moths captured while undertaking movement (Callahan et al., 1972; Greenbank et al., 1980) can provide additional information for inclusion in such a model, or, alternatively, as when the moths are found to be in a postteneral, prereproductive state (Johnson, 1960), may be interpreted as providing supporting evidence for identifying a moth population as migrants. Physiological and behavioral information can provide insights into the role of migration in a species' survival strategy (Gatehouse, 1987), and thus may make a worthwhile contribution to procedures for forecasting and managing migrant pests.

2. Meteorological and Climatic Observations

Insect migration is influenced by a wide variety of meteorological factors (Drake and Farrow, 1988). Movements on the between-region scale are wind-assisted (Fox, 1975, Drake et al., 1981; Farrow, 1984; Pedgley, 1985), and information on the winds at the time of flight is, therefore, of great value in the interpretation of observations and trap catches. The occurrence of an airflow capable of carrying the moths from a possible source region is strong supporting evidence for immigration (Hartstack et al., 1982; Raulston et al., 1982). Convergent airflows can concentrate airborne insects and may play a role in initiating outbreaks (Drake and Farrow, 1989).

Long-distance flights are made at heights between about 100 and 2000 m (Drake et al., 1981; Drake, 1984; Drake and Farrow, 1985), where nocturnal winds are approximately geostrophic, and usually much faster than the moths' airspeeds. Flight trajectories can be estimated from surface or 850-hPa synoptic charts, by backtracking the wind at the appropriate altitude (Scott and Achtemeier, 1987), and used to identify possible source regions (Riordan, 1979; Hartstack et al., 1982; Oku, 1983; Drake and Farrow, 1985). In the temperate zone, synoptic-scale weather features take only a few days to pass over a particular locality and the direction of the geostrophic wind often varies markedly from night to night. Identification of immigrants with particular wind systems will only be possible if the night on which the movement occurred is known, so traps should be cleared every day. In tropical regions, where wind systems may be somewhat less changeable, streamline charts should be used in the calculation of trajectories (Farrow, 1984).

Synoptic-scale analysis may be inadequate when the insects are flying in such large-scale boundary-layer wind systems as sea breezes (Drake, 1982b) or nocturnal low-level jets (Drake, 1985). Such situations may be identifiable by weather forecasters with experience of local conditions, or they can be detected by direct measurement of the upper winds, using pilot balloons (Meteorological Office, 1968). Detailed observations of windfields can be made with an aircraft equipped with a Doppler-radar windfinder, a technique that is of particular value when studying convergence (Rainey, 1976).

Flights made near the ground, which generally cover shorter distances, are subject to different weather influences. When weather conditions are settled, and generally favorable for moth flight, a surface temperature inversion often forms, and there may be strong wind shear at altitudes below about 200 m. Winds near the surface will be very light then, and the moths' flight boundary layer may extend to a height of 10 m or more; nevertheless, moth flight is often downwind (Brown, 1970; V.A. Drake and R.A. Farrow, unpublished). Winds within the shear zone cannot be estimated reliably either from synoptic charts or from measurements made at the usual height of 2 m; they are best determined by direct measurements, using instruments mounted on a tower. Short-distance movements may also be strongly affected by small-scale boundary-layer phenomena, especially storm outflows, that could carry insects over distances of the order of 10 km in a direction quite different from that of the general windflow (Schaefer, 1976; Pedgley, Reynolds, Riley, and Tucker, 1982). *Heliothis* movements on this scale are not consistently in the direction of the prevailing wind (Hendricks et al., 1973b).

Meteorological observations are also required for analysis of the development of candidate source populations. Emergence dates can be estimated from degree-day accumulations (Smith and McDonald, 1986). Daily temperature values recorded at the climatic observation stations of the national meteorological service are usually sufficient for this purpose, but may need to be modified to take account of the insulating effects of the host plant's vegetation cover and, during pupation, the soil (G.P. Fitt, personal communication). In arid regions, population levels may be related, via host plant quality, to rainfall (Farrow and McDonald, 1987). This may be patchy and not recorded adequately by available raingauges: direct observation of vegetation condition, for example by survey or by satellite remote-sensing (Riley, 1989) may then provide the most reliable indication of potential for insect production. Climate data (that is, averages over many seasons), when combined with a development model, could be used to predict the regions where adult emergence is likely to occur at any season. Information on prevailing winds may be used to determine the most probable migration trajectories at any season, and to indicate the role of population redistributions in regular seasonal patterns of infestation (Lin, Sun, Chen, and Chang, 1963; Brown et al., 1969; Wolfenbarger et al., 1973; Hartstack et al., 1982; Drake and Farrow, 1985).

3. Associated Migrants and Airborne Material

When weather conditions are favorable for wind-assisted migration, a number of species may undertake long-distance movements simultaneously (Fox, 1975; Drake and Farrow, 1985; Farrow, 1984). The sudden appearance of known migratory species may constitute supporting evidence that *Heliothis* appearing at the same time are recent immigrants (Morton et al., 1981; Farrow, 1984; Pedgley, 1985). The air currents on which the moths have been carried may also be indicated by such natural tracers as pollen, spores, and plumed seeds (Close, Moar, Tomlinson, and Lowe, 1978; Pedgley, 1980) or dust and smoke (Fox, 1969).

V. Application of the Methods

Each of the methods described above has particular strengths and weaknesses. Most are applicable only to movements of certain types. Many of the methods are complementary, and when employed simultaneously can produce much better evidence for movement than could be obtained by any method alone (Greenbank et al., 1980; Drake and Farrow, 1983; Riley et al., 1987). There are a number of ways in which complementary observations can be particularly valuable:

1. Evidence (for example, from trap catches) that moths have immigrated into a locality will be complemented by direct observations (for example, visual or radar) of the immigration flight. The direct observations could be made in the source or destination regions of the migration, or at intermediate points.

2. Direct observations of movement in progress will be complemented by capture of a sample of the insects undertaking the movement. This will provide both firm evidence of identity and information on the migrants' physiological condition.
3. Observations of immigrations and of movements in progress will be complemented by phenological observations of candidate source populations.
4. Almost all types of evidence for movement will be complemented by information on the meteorological and climatic conditions in which the movement occurred, and in which the source population developed.

Some features of these methods, and of the phenomena they are designed to detect, have implications for the organization of research on *Heliothis* movement (Kennedy and Way, 1979; Rabb, 1979; Farrow and Daly, 1987). The variety of methods, and their complementarity, points to the need for a multidisciplinary team approach in which insect ecologists combine with nonentomologists with specialist skills as various as palynology, radar engineering, and meteorology. The benefits of employing a number of methods simultaneously indicates that field work should generally be undertaken collaboratively. Intensive studies by a team of observers at a single field site may be required to produce information on movement that can be related to changes in local *Heliothis* populations. Such intensive studies are of necessity short term. Longer-term wide-area monitoring of populations will also be of value; for this, cooperation between regional governmental agencies, and collaboration between scientists based at widely separated locations, may be necessary. Observations should not be confined to crops and regions where *Heliothis* is of direct economic importance, because immigrant populations may have their origins elsewhere and because pest management in high-value crops is likely to suppress populations to levels where many observation methods are ineffective. Similarly, observation of immigration may be easiest at the periphery of the species' normal range, as in the middle of the range the net effect of immigration and emigration may be too small to detect (Phillips, 1979).

Acknowledgments. The traps shown in the figures were developed mainly by Dr. R.A. Farrow and Mr. J.E. Dowse. Dr. Farrow, Dr. G.P. Fitt, and Dr. R.D. Hughes read and made helpful criticisms of a draft.

Chapter 11

Methods for Studying Behavior

G.P. Fitt and G.S. Boyan

I. Introduction

This chapter deals with techniques for studying the behavior of adult and larval *Heliothis*, primarily those associated with the utilization of host plants. These behaviors form a continuum throughout the lifetime of the insect from mating behavior, to the location and selection of appropriate host plants by females and subsequent feeding activities of the larvae. We neither discuss techniques for quantifying adult movements involved in host location nor migratory movement. These are covered in Chapter 10. Most emphasis is placed on techniques for use with intact, active insects. The aim is to present a review of techniques that allow behaviors to be measured and interpreted in a way that is meaningful to the natural behavior of the insect. Electrophysiological techniques used to more precisely define responses to chemical stimuli, or the activity and specificity of particular receptors are discussed briefly (pp. 139–142), but such techniques have not yet been widely applied to *Heliothis* spp. and are well described elsewhere (see references pp. 139–142). As a model for the way field and laboratory studies of behavior and physiology can be integrated, we discuss predator avoidance in adult *Heliothis* and the auditory system involved in that behavior (pp. 142–150).

Heliothis are primarily crepuscular/nocturnal insects. Occasionally, adults may feed and oviposit during the day, particularly if the day is cloudy, but most activity occurs at dusk and during the night (Lingren et al., 1982; Topper, 1987b; Ramaswamy, 1988; Fitt, 1991). This makes direct observation of behavior difficult. Many aspects of host selection behavior and chemosensory physiology of *Heliothis* species remain poorly understood (Ramaswamy, 1988; Fitt, 1991), despite the pest status of the genus worldwide (Fitt, 1989). In contrast, the auditory system and predator avoidance behavior are paticularly well understood (pp. 142–150) and many of the principles of nervous system structure and function described in other noctuids are probably equally applicable to *Heliothis*.

Detailed studies of animal behavior usually proceed through three phases: 1) observation and description of the animals behavioral repertoire; (2) quantifying behavior using field measurements, bioassays, or electrophysiological techniques, and (3) relating findings to existing principles or theories. Here we deal first with techniques for studying mating behavior, then move on to examine the processes involved in host selection, oviposition, and finally predator evasion behavior.

II. Mating Behavior

Quantitative descriptions of *Heliothis* mating behavior are few. Teal et al. (1981a) have produced the only ethogram of reproductive behavior for *Heliothis* based on laboratory observations of *H. virescens*. Observations were made under low light or red light in a plexiglass wind tunnel (1.5 × 0.5 × 0.5 m) through which air was pulled at a constant rate. Males were released at the opposite end to five calling females and their behaviors were recorded on tape. The behavior of individual males in small plexiglass cages was monitored using video equipment. By identifying the major behavioral components and observing a series of mating encounters, they were able to calculate transition matrices between different behaviors and thus link the components into a sequence leading to copulation.

Most studies of *Heliothis* reproductive behavior have centered on the timing of mating activity in relation to other activities, on the response of males to female produced pheromone, and catches of males in pheromone baited traps. Temporal aspects of mating activity can be quantified by direct observation, either by collecting mating pairs in the field or recording activity at mating tables. Mating tables are simply a platform set up in the field, on which virgin females with their wings clipped are placed together with some foliage to provide shelter and resting sites. The tables can be observed continuously, or at intervals, to record the timing of female calling behavior and male responses. Features such as the length of time spent in copula could be obtained in this way. Mating tables have been used extensively in studies of the mating competitiveness of different strains of *H. virescens* (Proshold et al., 1983).

A number of studies of male responsiveness to pheromones have been conducted in the field (Rothschild, 1978; Sparks et al., 1979; Rothschild et al., 1982; Teal, Heath, Tumlinson, and McLaughlin, 1981b), where different individual components or blends were exposed to field populations in standardized traps and the number of males attracted was recorded (see Chapter 4 for a description of traps and the interpretation of catches). Such research has been particularly valuable in the development of attractive pheromone blends for a number of Lepidoptera (Roelofs, 1977), but it provides little insight into the behaviors involved in short- or long-range orientation and attraction to the odor source. More quantitative observations of male orientation to pheromone sources can be made using nightvision devices or low-light video systems (see below). David et al. (1982, 1983) describe the use of a tower-mounted video system to obtain two dimensional flight tracks of male, *Lymantria dispar* and *Spodoptera littoralis* approaching pheromone sources, while simultaneously recording the behavior of the plume itself. Similar video techniques have been applied to some noctuids (Murlis, Bettany, Kelley, and Martin, 1982) and have been used also to obtain three-dimensional flight trajectories of *Heliothis* flight behavior above crops at night (Riley et al., 1990; see Chapter 10). Perry and Wall (1986) review a series of studies in which the pattern of male catches in different arrangements of traps was used to model both the structure of the odor plume, and the flight of moths with respect to odor sources.

Studies of the specific behaviors involved in responses to pheromones may be best conducted in wind tunnels where the identity and concentration of olfactory stimuli, the structure of the plume, and the quality of the insect can all be controlled. By using video cameras in conjunction with wind tunnels, the sequence of behaviors involved in take-off, in following the odor plume in flight, and in eliciting closure with the source can all be quantified (Miller and Roelofs, 1978; Carde and Hagaman, 1979; Linn, Campbell, and Roeloffs, 1986). For example, such studies with three Lepidopterans, including the noctuid *Trichoplusia ni*, have shown that the complete blend of the multicomponent pheromone is involved in initiating upwind flight of male moths and contributing to pheromone specificity (Linn et al., 1986). Previously it was thought that only the major components stimulated the upwind approach of males, whereas the minor components stimulated close-range courtship activities.

The same seems true of *Heliothis* pheromones. All six aldehydes in the *H. virescens* pheromone influence mating success of males (Teal, Tumlinson, and Heather, 1986). Although individual components appeared to affect specific behaviors, in no case did a single compound elicit a specific behavior in the mating sequence. Studies of male responses to pheromones have provided much of what is known about insect orientation to odor plumes in general (David et al., 1983; Carde, 1984; Baker, 1986; Payne, Birch, and Kennedy, 1986).

III. Host Selection and Oviposition

1. Overview of Host-Related Behaviors in Adult *Heliothis*

Ramaswamy (1988) and Fitt (1991) provide comprehensive reviews of the behavior and sensory stimuli involved in host selection by Heliothinae. For most holometabolous insects such as *Heliothis* spp., larval survival is very much dependent on the appropriate responses by ovipositing females. Although partially grown larvae may be quite mobile and hence able to exercise some selection of feeding site, neonates and early instars are relatively restricted in their movement.

Generally, the host selection process of phytophagous insects is regarded as a catenary process or linked sequence of behaviors, often to different stimuli, in which each response in the chain brings the insect into the zone of influence of the next stimulus (Thorsteinson, 1960; Kennedy, 1965). The sequence thus involves a series of behavioral responses by the female to cues from the plant or its environment, that ultimately lead to the location of the host habitat, the location of the host plant, and its acceptance or rejection for oviposition (Kogan, 1977). These external stimuli may be excitatory or inhibitory and various combinations of chemical, physical or visual attributes may be used as proximate cues at each stage, even though a dominant "sign stimulus" (often a host-specific chemical) may appear to be the major factor leading to oviposition.

The view that host range is determined by a few key stimuli is now shifting to the view that insects respond to the summation of various stimuli from the plant (Dethier, 1971). These proximate cues may be perceived via a number of sensory modalities; gustation, olfaction, mechanoreception, and vision (Ramaswamy, 1988), each of which may vary in its sensitivity and specificity. Olfactory and visual cues are perceived from a distance and used in prealighting discrimination between plants, whereas gustatory and mechanical cues require contact with the plant and so are useful only in post-alighting discrimination. In addition, the physiological state of the insect may alter its sensitivity to external cues and, therefore, its thresholds for the performance of particular responses. For example, if acceptable hosts are scarce, host-seeking insects may become less discriminating with increasing time since the last successful oviposition (Singer, 1982; Fitt, 1986), but sometimes not, as for example, with cabbage rootflies (Nair and McEwen, 1976) and some species of *Dacus* that lay few eggs in the absence of preferred host plants (Fitt, 1986). As a result of reduced discrimination or specificity, females may accept plants of low ranking, that may or may not be suitable for larval development, but that are normally unacceptable in the presence of more highly ranked plants. It is the interaction between these external and internal excitatory and inhibitory stimuli that ultimately results in the selection and acceptance of a plant for oviposition or feeding (Miller and Strickler, 1984).

When considering the process of host selection in *Heliothis*, and in other highly mobile organisms, it is also important to recognize the interaction between their mobility and the behaviors involved in host location and acceptance. In many agricultural situations, host-seeking adults will be faced with large monospecific patches of particular plants. Local movement then becomes the only mechanism by which adults can express preferences for particular plants. Using cues derived from the plants occurring in the local area, the insect can either remain and deposit eggs, or move elsewhere in search of perhaps more suitable hosts. Females will be confronted only rarely with a simultaneous choice between alternatives, and the range of hosts utilized will be as much a reflection of a hierarchy of host acceptance, as they are of preferences during host location. In general, there has been little study of the behaviors involved in the location of host patches by *Heliothis* spp. (Fitt, 1991).

Most of the research techniques that follow relate to the final stages of host selection after the insect is within the host habitat (host recognition and discrimination) and to host acceptance, as signalled by oviposition. Ramaswamy (1988) provides one of the few detailed descriptions of short-range host selection and oviposition behavior in *Heliothis* spp. from which he was able to construct an "ethogram" of the behavioral chain leading to oviposition in *H. virescens*. He describes 10 distinct behaviors involved in egg laying and the transition frequencies between them. Oviposition occurred in 2–4 major bouts, each consisting of 3–8 minor bouts in which 5–10 eggs were deposited. The sequence of behaviors during a bout included antennal movement, wing buzzing, walking, wing fanning, flying, abdomen bending, ovipositor drag, egg deposition, and abdomen pulsate (which usually signalled the end of a major bout). Not all behaviors were represented in a sequence and there were many reversions to previous behaviors. Similar ethograms have been prepared for some butterflies. For example, in the cabbage white *Pieris brassicae*, females show the following sequence in response to plants: approach, landing, drumming with the foretarsi, curving of the abdomen, touching the leaf surface with the extended ovipositor, and finally oviposition (Klijnstra, 1982). However, to quantify these components of the behavioral sequence and the roles of different stimuli in eliciting particular responses, it is usually necessary to conduct controlled experiments or bioassays.

2. General Considerations for Conducting Bioassays

Useful discussions on the design of bioassays for particular behavioral studies can be found in Kennedy (1977), Finch (1980, 1986), Baker and Carde (1984), and Miller and Miller (1986). In designing experiments or observations to quantify the behavioral responses involved in host selection, it is important to consider the stage of the host selection sequence under study and design assays appropriate for each. There are a number of factors to be considered in designing an appropriate bioassay; for example, arenas with and without airflow, using real plants or artificial substrates, recording activity of the insects or only the numbers of eggs laid. The type of assay required will depend on the organism and the precise objectives of the study. For example, to examine the sequence of host selection behavior in *Heliothis* adequately it may be necessary to allow as natural a sequence of flight behavior as possible, whereas particular aspects of host acceptance may be addressed with suitable laboratory assays (Kennedy, 1977).

There are a number of requirements that must be met for all behavioral bioassays. First, there must be no bias in the movement of insects within the arena or in the distribution of eggs due to factors other than the test plants or substrates. Thus in laboratory bioassays, standardized illumination, temperature, humidity, and air movements are essential. Such control is clearly not possible in the field. Position effects should be minimized, in both laboratory and outdoor enclosures, by rotating alternatives among positions of a grid at regular intervals during the test period or between replicate runs. Alternatively, the choices may be arranged on a slowly rotating turntable (Finch, 1980), such that their positions relative to external stimuli are changing constantly. Second, in order to obtain reproducible responses, the responsiveness or "internal state" of the test insects must be standardized. In bioassays of host plant chemicals it should be mandatory that the test insects be used only once, when they are at the most responsive age, usually just prior to the start of oviposition (Finch, 1980).

Unfortunately, many of the bioassays that have been used with *Heliothis* to identify attractants or oviposition stimulants do not discriminate between the sensory modalities involved, for example between short-range olfaction and contact chemoreception. The following sections describe a range of laboratory techniques that may be used to identify the cues involved in host selection and then discuss the design of bioassays to quantify host preferences and specificity.

3. Prealighting Olfactory Stimuli

A. *Olfactometers*

The most widely used technique to examine the responses of females to particular host-derived olfactory stimuli is the olfactometer. These can be of several designs, the simplest being a *y*-tube olfactometer in which the insect is presented with a choice of two airstreams, one carrying a plant odor or particular volatile and the other clean air. A response is recorded by the movement of the insect towards the source of one airstream or else on the basis of eggs deposited on a suitable surface (e.g., gauze) placed across the entrance port of each airstream.

Alternatively, an olfactometer may consist of a simple tube or extended cage along which a gradient of host odor or specific volatiles is established. This can be achieved by placing potted plants outside the screened end of a chamber such that volatiles can pass inside. Such simple olfactometers have been used for choice experiments by placing alternative plants or volatiles at each end; preference being expressed in terms of the distribution of insects along the gradient or the relative numbers of eggs laid at either end (e.g., Rembold and Tober, 1987). The interpretation of such crude bioassays is almost impossible, particularly if no airflow has been used. First, if preference is based solely on the numbers of eggs laid on an unnatural surface, the assay will reveal only attractants that also elicit oviposition in conjunction with a physically suitable substrate. Volatiles involved only in attraction would not be detected. Second, if potted plants or other plant material behind a screen are used, it may be difficult to exclude the role of visual stimuli.

Another problem with olfactometer studies in general is the difficulty of relating the results to the process of locating a distant odor source in the field (Kennedy, 1977). The types of responses shown in the apparatus may not occur under more natural conditions, as the steep gradients in odor concentration used may permit reactions that would only occur close to the source in the field. For example, some insects respond to different stimuli when walking rather than when flying (Kennedy, 1977). An olfactometer for *Heliothis*, therefore, must be sufficiently large to allow flight. In the absence of detailed observations of behavior within the apparatus, most simple olfactometers provide undiscriminating assays in that they measure the net result of several behaviors (ortho- or klinokineses or chemotaxes, see Dethier, Barton-Browne, and Smith (1960) and Kennedy (1986) for a definition of terms). Kennedy (1977) provides a discussion of the attributes of different assay techniques and describes more discriminating techniques that allow the measurement of one type of response to a particular stimulus.

B. Wind Tunnels

Much experimental evidence suggests that chemically mediated, optomotor-guided anemotaxis is the principal mechanism of long-distance orientation to chemical stimuli by flying insects (David et al., 1982, 1983; Carde, 1984). The typical flight track of moths orienting along an odor plume is a series of zigzags or lateral reversals along an upwind trajectory, that eventually lead to location of the odor source. Downwind casting may also occur if the moth loses contact with the plume by overshooting the source (Kennedy, 1977). Useful introductions to the orientation of insects to olfactory stimuli can be found in Carde (1984), Bell (1984), and Payne et al. (1986). In the case of *Heliothis* there has been little study of the involvement of up-wind anemotactic flight in response to host volatiles by females (but see Ramaswamy, 1988), thus most of the examples given here refer to male responses to pheromone. Nevertheless, the experimental techniques are similar and their use to quantify responses of *Heliothis* spp. to host volatiles should be expanded greatly.

The best method for detailed study of olfactory responses is with the use of low-speed wind tunnels, in which quantitative analyses of flights and orientation mechanisms is possible (Kennedy and Marsh, 1974; Finch, 1980; Baker and Carde, 1984), by the use of controlled lighting and video equipment. A dispenser impregnated with the volatile (pheromone or host volatile) is placed at the upwind end of the wind tunnel and the responses of insects placed at the opposite end are observed and if possible quantified. Among Lepidoptera, this technique has been applied mostly to studies of orientation to pheromone plumes (Marsh, Kennedy, and Ludlow, 1978; Miller and Roelofs, 1978; Haynes and Baker, 1989), but is equally applicable to responses to plant-derived stimuli (Finch, 1980; Haynes and Baker, 1989). Most studies involving wind tunnels have attempted to rigorously control the pattern of air movement, usually achieving laminar flow. Hence, the nature of the odor plume was controlled and simplified. Kennedy, (1986) points out some of the dangers of such an approach in light of the findings of Wall and others (Perry and Wall, 1986), that demonstrate the disruptive and diffusive effects of plant canopies on airflow and odor movement.

4. Prealighting Visual Stimuli

Prokopy and Owen (1983) define three properties of plants that may serve as visual cues to foraging insects: (1) spectral qualities (hue, saturation, intensity); (2) dimensions; and (3) patterns. Spectral quality is more likely involved in the host discrimination of day-flying foragers than of nocturnal *Heliothis*.

Most studies of *Heliothis* host selection suggest specific visual cues are not of major importance (Schneider et al., 1986; Ramaswamy, 1988; Fitt, 1991) and they are not dealt with exhaustively here. By using a simple laboratory bioassay, Callahan (1957) nevertheless demonstrated discrimination between different wavelengths of low, equal-intensity light by *H. zea* indicating that these insects could perceive hue. Oviposition preference decreased with increasing wavelength from blue to red. The spectral sensitivity of the compound eyes of *H. zea* and *H. virescens* is identical (Agee, 1973) and color vision may play a minor role in foraging activities at low-light intensities around dusk. Some studies of male responses to pheromone traps (Gross, Carpenter, and Sparkes, 1983) have noted the visual acuity of *Heliothis* adults and indicate that general visual characteristics of plants (shape, outline, and height in relation to surroundings), which comprise their dimension and pattern, may well play a part in short-distance discrimination prior to landing. Some techniques for defining the role of the visual characteristics of plants in host selection are available in the above papers and in Prokopy and Owen (1983).

5. Postalighting Stimuli

Postalighting discrimination is achieved most often by chemoreceptors in response to contact chemical or very short-range olfactory cues, or else by mechanoreceptors in response to leaf surface texture or other physical characters (e.g., presence or absence of trichomes). Identifying the cues responsible for postalighting discrimination can also be achieved using artificial substrates in laboratory assays. These may be artificial leaves made of filter paper or cellulose sponge (Feeny, Rosenberry, and Carter, 1983) or simply sheets of absorbent paper impregnated with particular chemicals of interest (Tingle and Mitchell, 1984; Jackson, Severnson, Johnson, Chapling, and Stephenson, 1984). Artificial leaves can be impregnated with selected plant extracts and responses observed or measured from rates of egg deposition. In defining the attractiveness of particular chemical stimuli, one substrate is usually impregnated with the test compound dissolved in a suitable solvent, whereas the other is treated with the solvent alone. Similar alternatives can be used to examine the effects of physical differences in plant surfaces. Cullen (1969) considered that surface chemistry, the presence of adult food, and high humidity were all less important to *H. punctigera* than surface texture in the selection of a particular oviposition site within the plant. He offered females a choice of seven different real and artificial oviposition sites, that varied in surface characteristics and showed that villous or rugous surfaces received considerably more eggs than did smooth surfaces. By using a simple laboratory experiment, in which he varied the surfaces contacted by the tarsi and the ovipositor, he was also able to demonstrate that tarsal mechanoreceptors were primarily involved in this discrimination. Reductions in the number of eggs deposited on glabrous cotton genotypes (Schuster and Anderson, 1976; Fitt, 1987) offers further evidence of this factor.

6. Host Preference

Singer (1986) gives a thorough treatment of the definition and measurement of oviposition preference. The term *preference* implies active choice from among a range of options, all of which are perceived simultaneously or in sequence, if learning is invoked (Miller and Strickler, 1984; Singer, 1986; Papaj and Prokopy, 1989). Choice necessarily implies that the insects possess the sensory characteristics needed to differentiate between different plants. Generally, bioassays of preference behavior are conducted in an arena where the insect is offered a simultaneous choice of alternatives. Most often this type of bioassay is used to examine the behavior of a group of insects from the same source and treated in the same way. Only rarely are they used to examine the behavioral preferences of individuals (Traynier, 1979). Arenas may vary from small cloth or perspex cages (approx. $1 \times 1 \times 1$ m) set up in the laboratory where the environment can be controlled completely (Traynier, 1979; Renwick and Radke, 1985; Feeny et al., 1983), to large field enclosures covering several square meters where potted or naturally growing plants are offered in different arrays (Cullen, 1969; Firempong and Zalucki, 1990a,b). The alternatives may be natural host plants, either of different species of plants, cultivars of the same species having different characteristics, or plants of the same species and genotype that have been treated in different ways (for example, by the application of chemical extracts). Alternatively, the choice may be between artificial substrates that have been treated in different ways; that is, treated with different chemicals to examine the role of suspected olfactory or contact chemical cues, treated with different colors to examine visual preferences, or created to consist of a different texture in order to examine the role of surface characters.

In their simplest form, choice experiments provide the test insects with only two alternatives, though multiple choices can be offered (Johnson et al., 1975). The results of such tests are most often derived solely from counting the numbers of eggs applied to the differing alternatives after a set experimental period (Johnson et al., 1975; Gross, 1984; Jackson et al., 1984; Hillhouse and Pitre, 1976; Martin, Lingren, and Greene, 1976). Differences in the proportion of eggs laid on the alternatives are then expressed as an index of preference (or deterrence):

$$\text{Preference index} = \frac{(B - A) \times 100}{(A + B)}$$

where A and B are the numbers of eggs laid on the alternatives (Renwick and Radke, 1985). Although this approach is adequate for many cases, the reliance on simply counting the numbers of eggs is unfortunate. It is highly desirable to combine choice tests with detailed observations of the insects during the experiment. Data collected from such observations can provide considerable insight into the mechanisms underlying the patterns of host preference or discrimination and may lead to more precise assays of pre- and postalighting host selection mechanisms (as described above). Such observations may also indicate whether the distribution of eggs in the test can realistically be applied to the field. Thus in the case of *Heliothis* one could measure the number of approaches to alternative plants, the number of landings and the length of visits to plants in addition to the final numbers of eggs laid on each plant (Fitt, 1987). van Lenteren et al. (1978) discuss a number of examples of host discrimination by parasitoids where measurements of the pattern of egg distribution alone wrongly suggested that females did not discriminate between parasitized and nonparasitized hosts, whereas observations of the rate of encounter with alternatives, showed that indeed they did discriminate.

Laboratory choice experiments have proven extremely useful in determining relative preferences in response to short-range or contact stimuli and have been used to evaluate the behavioral rankings of different hosts by some *Heliothis* species (Johnson, Stinner, and Rabb, 1975; Martin et al., 1976; Hillhouse and Pitre, 1976; Firempong, 1987). However, their interpretation and extrapolation to the field requires considerable care, particularly where host rankings have been assigned solely on the basis of differences in the numbers of eggs deposited on the alternatives. Miller and Strickler (1984) provide a clarifying discussion of the terms normally applied to such studies. The terms *selection* and *preference* imply the presence of a number of alternatives each of which can be assessed, whereas *acceptance* does not specify whether or not alternatives are assessed but simply indicates that consummatory behaviors occur (Thorsteinson, 1960). The difficulty in applying results from laboratory choice tests to the field for such insects as *Heliothis*, is that usually females will not have a choice once colonization of the patch has occurred, particularly when they are exploiting large monospecific patches of cultivated crops. They must accept or reject a plant on the basis of cues from that plant alone, unless learning or the recall from memory of details of plants sampled earlier is invoked. Thus Thorsteinson (1960) concluded that "foraging by herbivorous insects is more akin to a series of take it or leave it situations (with the durations of staying [in a patch] being variable), than it is to comparison shopping." Because of limitations imposed by temporal or spatial patterns of host availability, the behavioral preferences or rankings revealed by choice experiments in the laboratory may not be reflected in the field.

No-choice experiments, in which insects are confined with particular plant species, can be used to determine the range of plants acceptable to a particular species. However, during such assays it is important to consider the influence of the insect's changing physiological state and hence greater propensity to oviposit on normally unacceptable plants after a period of deprivation. This effect must be controlled by direct observation, or by measurement of the time required before different plants are accepted (see below). This approach is used widely during studies of potential biological control agents for the control of noxious plants where researchers must be sure which plants may be accepted under any conditions.

By manipulating the cues available, choice experiments may also be used to determine the relative importance of stimuli derived from different plants (chemistry, physical features) in eliciting components of host selection and acceptance. Such studies are discussed more fully below.

7. Host Specificity

Host preferences revealed by choice experiments are not necessarily fixed. They can vary within an insect species or within an individual in response to changes in internal physiological state caused by a shortage of preferred hosts (Wiklund, 1981; Singer, 1982; Roitberg and Prokopy, 1983), or previous experience with hosts (Traynier, 1979; Rausher, 1980; Papaj and Rausher, 1983). When oviposition behavior is not consummated, a change in threshold for particular responses to host stimuli may result from changes in the peripheral (Davis, 1984) or central nervous system (Dethier, 1976; Singer, 1982). Consequently, in the absence of their preferred host, specialized insects may become less specific in their distribution of eggs. In such polyphagous insects as *Heliothis*, which nevertheless display a hierarchy of preferences, the pattern of host utilization within a locality may also change according to the relative abundance or apparency of potential hosts (Fox and Morrow, 1981; Wiklund, 1981).

One technique to quantify the specificity of host plant choice by individual females, and to measure any changes due to experience, is to manipulate their encounters with plants and record the pattern of acceptance. Singer, (1986) describes the technique in detail. Briefly, the experimenter holds the insect by the wings, or perhaps on with a holder affixed to the thorax, and then manipulates the sequence and timing of encounters with different plants. This can be done in the laboratory (Singer, 1982) or in the field (Singer, 1983). The female is placed on the leaf surface (or perhaps on some other plant structure) so that chemo- and/or mechanoreceptors on the tarsi, antennae, or ovipositor come into contact with the plant. Acceptance is recorded if the characteristic abdomen curling and extrusion of the ovipositor that precede oviposition are observed. If the insect is then removed from the plant before oviposition occurs, the interval between ovipositions can be prolonged and any change in the range of acceptable plants during this interval can be determined. An individual in which the range of acceptable plants does not change rapidly when deprived of oviposition is probably more specialised behaviorally speaking, than one for which the range of acceptable plants increases faster. For *Heliothis* this technique should be most useful in showing the rank order of acceptability of different hosts. Moreover, because the preferences of individual insects can be measured, questions related to the level of intraspecific or intrapopulation variability in host preference can be addressed. This variability may be masked by techniques that examine the collective responses of several females. Singer (1982) showed that in *Euphydryas editha* host acceptance or rejection following such manipulation was correlated strongly with that shown on the same individual plants by freely alighting insects. Although such manipulative experiments have not been done with *Heliothis*, preliminary attempts (M.P. Zalucki, personal communication) suggest it is feasible if the moth is tethered. Prokopy, Averill, Cooley, and Roitberg (1982) describe another method of manipulating encounters with different hosts in order to determine whether host acceptance by female Tephritids is influenced by prior experience. These techniques may also be applicable to *Heliothis*. Another technique used to determine the ranking of host plants is to offer a choice of plants over a number of days to the same insect(s), each day removing the most preferred plant of the previous day. Thus the number of choices is reduced over time and the most preferred plant each day gives a ranking of hosts. There are some limitations for this method, because not only is the range of plants changing with time, but the test insects are aging. Maximal egg maturation and oviposition occurs over a period of only 3 to 5 days in *Heliothis*. Females may become less discriminating as they age or as their complement of eggs is reduced. Thus the range of plants initially offered should not be great (4–5) because rankings late in the assay may be unreliable. Manipulation of larval diets and/or adult experience with different plants (Jermy, Hanson, and Dethier, 1968; Hanson, 1976; Traynier, 1979) can also be used to examine the possibility of induced or learned preferences.

8. Field Observations of Behavior

Observation of host selection and oviposition behavior in the field has some advantages over laboratory studies, in that the entire sequence of behaviors may be studied in a totally natural environment. However, it has the disadvantage that the relative importance of specific cues cannot be assigned easily. Field studies should serve as the basis for more precise and intensive laboratory or field cage studies.

One direct method for quantifying host plant choice and preferences in the field is the "flight path sampling technique" described by Stanton (1982). Jones (1977) describes a similar technique for analyzing flight patterns within artificially established grids of host plants with differing characteristics. Individual ovipositing females are observed and followed for as long as possible and details of visits to plants are recorded on a portable tape recorder. Each visit to a plant can be timed and the plant is marked with sequentially numbered flags so that the flight track can be mapped later and analyzed to show the average distance between plant visits, turning angle between visits, and so on. In areas of native vegetation or weedy areas where a range of potential hosts is available, the preferences of individual females to alight on, or oviposit on, different plants can be quantified by measuring the diversity of plants along the flight track using quadrats. If plants are visited in proportion to their abundance along the track, then females are probably visiting them at random. In addition, by comparing the diversity of plants encountered along flight paths with that from random transects through the study area, it is possible to show whether the flight path allows a random encounter rate with different hosts. Discrimination among conspecific plants can also be quantified in this way by measuring characteristics of the plants alighted on by females. If females show postalighting discrimination among plants, the characteristics of hosts they alight on and accept should differ from the characteristics of hosts that they alight on but reject (Stanton, 1982; Mackay, 1985; Mackay and Jones, 1989; Oyeyele and Zalucki, 1990; Zalucki, Brower, and Malcolm, 1990). Zalucki et al. (1980) describe another technique for tracking the flight of day-flying butterflies through patches of host plants. They used two fixed recording theodolites to measure the coordinates of a moving butterfly at intervals along the flight path. An ALGOL program was then employed to translate the angular readings into (x,y,z) coordinates for each fix. The technique was used to study flight patterns of *Danaus* within and between patches of milkweed (Zalucki and Kitching, 1982a).

The nocturnal habits of *Heliothis* spp. make them particularly difficult subjects for quantitative studies of oviposition behavior, unless observations are restricted to dusk. However, the techniques described above may be applicable with the careful use of headlamps or a combination of infrared lighting and nightvision devices (Lingren et al., 1982), or near infrared video equipment (Riley et al., 1990). Visible light headlamps can be used to collect some behavioral information (e.g., collecting mating pairs at different times through the night) but, unless used carefully, the moths are disturbed easily. Nightvision goggles (NVGs) offer the best alternative though they are expensive. These are low-light, image-intensifying devices that in combination with infrared spotlights or headlamps have been used in a number of studies on *Heliothis* behavior (Lingren et al., 1978b; Lingren and Wolf, 1982; Fitt, 1991; G.P. Fitt, R.A. Farrow and J.R. Raulston, unpublished).

Topper (1987b) inferred various behaviors; movements in search of nectar and oviposition sites, in populations of *H. armigera* in the Sudan from differences in densities of moths observed in different parts of a diverse cropping system during the day and night. He estimated the daytime distribution of adults from flush counts in various crops during the day and the night time distribution from counts using torches. Similarly, Lingren et al. (1982) and G.P. Fitt, R.A. Farrow, and J.R. Raulston (unpublished) used headlamps and nets to capture adult *Heliothis* in cotton crops throughout the night. The behavior (sitting, flying, feeding, ovipositing, mating, hovering) of each moth was recorded allowing a time profile of behaviors to be developed. Such methods give little information on the cues involved in different activities but provide the basis for formulating questions that might be addressed.

Nightvision goggles have been used to quantify densities of adults flying within and at the boundaries of different crops (G.P. Fitt and R.A. Farrow, unpublished), from which inferences can be made about their relative attractiveness and the means by which moths are held within patches after colonization (Fitt, 1991). NVGs have also been used extensively to quantify the temporal pattern of responses by male *Heliothis* to pheromone traps or calling females (Lingren et al., 1978b; G.P. Fitt, unpublished) and to estimate trap efficiency. Individuals can be observed easily with NVG's and it may be possible to track the movements of ovipositing females between plants, though observations of specific oviposition behaviors on plants are difficult (G.P. Fitt, personal observation). No such studies have yet been attempted.

It is also possible to study flights in response to olfactory stimuli in the field. Murlis et al. (1982) describe a technique for analyzing moth flights towards pheromone sources using image intensifiers and cine camera to follow male *Spodoptera littoralis* moving along a plume that was floodlit with infrared light. They describe in detail the types of measurements that are possible using frame-by-frame analysis of the movie film. Murlis (1986) also describes techniques for characterizing the form of odor plumes. These involve either the use of chemical tracers added to the plumes in conjunction with fast-flame photometric detectors (Legg, Strange, Wall, and Perry, 1980) or the use of ionized air as a tracer and its detection electrically. This technique can be applied in wind tunnels (Boreham and Harvey, 1984) or in the field (Legg et al., 1980; Murlis, 1986) and has shown that odor plumes emanating from a fixed source consist of intermittent short bursts of odor when perceived from a fixed point downwind. The intermittency is due to imperfect mixing within the plume, undulation and meandering of the plume and synoptic scale changes of wind direction (Murlis, 1986). An analysis of intermittency functions of ion signals during studies with moth pheromones gave a modal burst length of 100 ms (range: 2 ms–1 s) and a return interval ranging from 30 ms to 30 s with a mode of 0.5 s. Both characteristics may be important in orientation. Moreover Legg et al. (1980) demonstrated that crops or other vegetation along the plume had a marked effect on the structure and chemical properties of the plume.

9. Field Experiments

Some information about oviposition preferences can be inferred by counting the numbers of eggs deposited in relatively small field plots, where the insect is assumed to make a choice among the alternatives. The same difficulties of interpretation that apply to laboratory choice experiments also apply here. The alternatives may be different species, different genotypes of the same species, or plots of the same species treated in different ways. Various statistically valid arrangements of plots or treatments can be used to overcome variation among the alternatives caused by extrinsic factors (e.g., soil types). These include randomized complete blocks, latin and lattice squares, and so forth (Cochran and Cox, 1957), each of which requires appropriate statistical treatment of the derived variables.

It is often possible to set up a replicated factorial experiment by using isogenic lines of crop plants that differ in only one or two characters and thus infer which physical or chemical factors or combinations of factors have the most influence on the abundance or distribution of eggs. As a result of field experiments, it may be possible then to design more controlled experiments to examine the behavioral responses to these factors in more detail. This type of study is common in research on host plant resistance and has been used to examine the pattern of oviposition by *Heliothis* on cultivars of a wide range of crops (Lateef, 1985; Lukefahr, Houghtaling, and Cruhm, 1975; Southern Coop. Series Bull., 1983). However, without direct observations of the insects it remains unclear whether females actually perceive a choice.

IV. Larval Behavior

Although in most holometabolous insects early instar larvae have little ability to select their foodplants actively, many nevertheless have the ability to detect, orient toward, and discriminate between different plants on the basis of short-range olfactory stimuli, chemical and physical stimuli after contact, or gustatory stimuli after a "test bite". Feeding behavior and sensory physiology of lepidopterous larvae, including those of noctuids, has been studied extensively (Schoonhoven, 1987). Hanson (1983) and Lewis and van Emden (1986) provide excellent coverage of the techniques used in quantifying insect feeding behavior, particularly for lepidopterous larvae (Hanson, 1983), and identifying some of the chemical stimuli involved. Hanson (1983) lists 3 distinct phases in host plant selection by larvae: (1) attraction to a potential food plant; (2) arrestment or cessation of movement; and (3) stimulation (or deterrence) of feeding.

Orientation of larvae towards host plants or extracts can be quantified using "small arena bioassays" (Finch, 1980) or "cross track-tests" (Hanson, 1983; Saxena and Rembold, 1984) in which the larva is constrained to move horizontally along a pathway in the shape of a cross. Test substances, or controls, are placed in vials at the ends of two arms (A and B) with airflow down each arm directed at the intersection point and with the larva placed at one end of the transverse arm. Alternatively, leaf disks or impregnated filter paper disks are placed a short distance (1 cm) on either side of the intersection (Saxena, Khattar, and Goyal, 1976). The larva then proceeds to the intersection where it may continue forward (no preference) or may turn left or right toward one of the choices. Choices can be interchanged between positions A and B between replicates. The whole arena should be screened from other external stimuli and illuminated uniformly. In any such bioassay it is essential that larvae are able to perceive both alternatives if a real choice is to be made, hence airflow within the arena is critical to the interpretation of the results. The proportion of larvae turning toward one or other choice can be expressed as an index, that is a measure of the attractiveness of A relative to B and can be analyzed with binomial statistics.

Although such assays may demonstrate the role of distance olfaction in orientation or attraction, this does not appear to be as important in food-plant selection as are gustatory signals derived from test bites (Woodhead, 1982) once larvae are on the plant. Moreover, the regulation of larval feeding may also be determined by such stimuli (Hanson, 1983).

Feeding preferences of lepidopteran larvae are examined most often using a leaf disc test, in which discs of material from 2 or more different host plants or genotypes are arranged in a grid pattern or circle within an arena, to which groups or single larvae are introduced. Preferences are recorded by measuring the relative numbers of larvae feeding on different discs at intervals after the start of the experiment or by measuring the quantity of leaf tissue consumed after feeding is stopped. The distribution of larvae or damage is then compared statistically with that expected from a random distribution. Leaf disc tests have been used to characterize the responses of larvae to phagostimulants or antifeedants (Blaney and Simmonds, 1988). Leaf discs or inert substrates (Lewis and van Emden, 1986) are treated individually with known concentrations of test compounds and exposed to larvae for a fixed period of feeding. The proportion of leaf area consumed, relative to discs treated with a control substance gives an index of the stimulatory or deterrent properties of the test substance (Lewis and van Emden, 1986; Blaney and Simmonds, 1988).

The acceptability of different foodplants and their impact on larval growth is assessed most often with no-choice tests, where the larvae are confined to feed in the laboratory or field on a single substrate. Various nutritional indices (food consumption, weight gain, efficiency of conversion, digestibility, etc) are then derived from measurements of larvae, plant material ingested, and excreta. Kogan (1986) discusses such techniques and measurements in detail.

In all behavioral assays with larvae it is important that the previous feeding history of larvae be standardized, as feeding behavior can be altered by experiential factors such as induction and aversion learning. Induction is a process whereby ingestion of a particular food for a period creates a preference for that food over other equally or more acceptable foods (Dethier, 1982). Induction of larval feeding behavior has been demonstrated in many species, including *H. zea* (Jermy et al., 1968; Hanson, 1976). Aversion learning has been demonstrated in some polyphagous Lepidoptera (Dethier, 1980b) and Orthoptera (Jermy, Bernays, and Szentezi, 1982) and occurs when larvae avoid host plants that are toxic or otherwise unpalatable after experience with them.

The role of particular antennal or maxillary receptors in stimulating or inhibiting larval feeding has been investigated by ablation experiments with many Lepidopterans (Chapman, 1974; Hanson, 1983; Schoonhoven, 1987). For example, removal of maxillary palps from larvae of *Sphinx ligustri* resulted in feeding on host plants that were normally rejected. More sophisticated electrophysiological techniques are discussed later. Dispersal of larvae within plants can be studied by releasing larvae at a particular position and tracking their movements by continuous or intermittent observations. There have been few such studies. Movements or preferences for particular plant parts can be inferred by analyzing the distribution of feeding individuals among these parts through time (Wilson et al., 1980). Wilson and Waite (1982) show how feeding preference coefficients can be calculated from distributional data. Measurements of feeding duration and numbers and ages of fruits attacked can be obtained by routine observations.

Many studies of host-plant resistance of crops to *Heliothis* use evidence of larval feeding on particular plant parts in the rating system. For example, Rogers (1982) discusses resistance in legumes based on the percentage of pods damaged. These studies rarely investigate larval behavior in detail.

V. Sensory Receptors and Electrophysiology

The purpose of this section is not to give an exhaustive treatment of *Heliothis* sensory physiology and electrophysiological techniques, but to provide a brief introduction to the relevant literature on this specialized area of behavioral research. Useful introductions can be found in Frazier and Hanson (1986), Mustaparta (1984), Payne et al. (1986), Stadler (1982,1984) and Zacharuk (1980,1984). Studies of insect sensory physiology require specialized equipment and expertise. Nevertheless, such studies are most valuable when they proceed hand in hand with behavioral studies. For example, Oldberg (1983) provides an excellent integrated study of pheromone responses in male *Bombyx mori*, relating the responses of interneurons in the ventral nerve cord to olfactory, visual, and mechanosensory responses to show how antennal movements and flip-flopping interneurons are involved in pheromone-directed turns in the upwind approach of male to female. The study of receptor responses totally in isolation from the natural behavioral repertoire of the animal should be avoided (below). Too often the behavioral significance of chemical or other stimuli has not been investigated in physiological studies of insects (Mustaparta, 1984).

Heliothis adults and larvae perceive their environment via a range of sensory receptors, though the sensory physiology of these species has been poorly studied to date. Receptors in the adult include olfactory chemoreceptors on the antennae (Stadler, 1984; Mustaparta, 1984; Blaney and Simmonds, 1988), contact chemoreceptors on the antennae, ovipositor, tarsi, and proboscis (Stadler, 1984; Ramaswamy, 1988) and mechanoreceptors on various parts of the body (Chapman, 1982), sometimes associated with chemoreceptors (Stadler, 1984). The morphology and function of insect chemosensilla and accessory structures have been reviewed by Zacharuk (1980,1984). Antennal chemoreceptors of moths include the hair-shaped sensilla, that is, sensilla basiconica involved in contact and odor perception,

and sensilla trichodea involved more specifically in detection of pheromones. In some adult insects the same olfactory cells are responsive to host
odors and to pheromones (den Otter et al., 1978; Dickens, Gutman, Payne,
Ryker, and Rudinski, 1984). Olfactory and/or gustatory chemoreceptors of
larval lepidoptera occur on the antennae and mouthparts (Hanson, 1983;
Stadler, 1984). Each antenna has three sensilla basiconica, while each maxil-
la has two sensilla styloconica on the galea and eight sensilla basiconica on
the maxillary palp (Schoonhoven, 1987). Larvae of many Lepidoptera also
have a gustatory epipharyngeal organ located on the epipharynx. Ablation
experiments in which different groups of these receptors were inactivated
have shown the sensilla styloconica and epipharyngeal organs to be most
important in distinguishing host from nonhost plants (refs. in Schoonhoven,
1987).

Several techniques have been developed for electrophysiological studies,
depending on the type of receptor under study (Kaissling, 1974; Stadler,
1982,1984; Mustaparta, 1984; Frazier and Hanson, 1986). The simplest
technique to determine the role of specific sensory inputs is to eliminate the
sensory organs (ablation). Physical removal of specific sensilla or receptor-
bearing organs, or acid treatments to cauterize the receptors (Stadler, 1984;
Ramaswamy, Ma, and Baker, 1987), can indicate the importance of different
modalities in eliciting a response. For example, Ramaswamy et al. (1987)
used ablation techniques to show that surface chemicals were perceived
largely by tarsal chemoreceptors in *H. virescens*. However, the technique is
valid only if appropriately treated control or sham operated insects are also
tested and if the treatment does not interfere with normal behavior. For
many, (e.g., proboscis, ovipositor) this is clearly not the case. More sophisticated electrophysiological techniques depend on the measurement of action
potentials produced in sensory organs (e.g., antennae), individual receptors
or single receptor cells, or the nerves emanating from the organ in response
to a stimulus. An apparatus for micromanipulation of electrodes and for the
delivery of the stimulus to specific receptors, and instruments to amplify
and display the electrical signal are required (Frazier and Hanson, 1986).
These techniques allow the sensitivity and specificity of receptor responses
to be determined by examining the relationship between stimulus, receptor
potential, and impulse activity. Analyses of the frequency of action potentials, and the intensity and shape of receptor potentials, increasingly by
means of computers (Hanson, 1983; Frazier and Hanson, 1986) can indicate
differences in the response to differing stimuli.

Schneider (1957) performed the first recordings from olfactory receptors
in the form of the summated receptor potentials of all the excited chemosensory cells on whole antennae, the electroantennogram (EAG). EAGs have
been used extensively since then in the identification of pheromones (Hummel and Miller, 1984), but less so for examining responses to volatile host
chemicals (Frazier and Hanson, 1986), and hardly at all with *Heliothis*

species. Whole antennae are prepared with a recording electrode on or penetrating the antennal surface and a reference electrode in the base of the antenna. The glass capillary electrodes are filled with an electrolyte solution (e.g., 0.1 M KCl) and connected to recording instruments via Ag-AgCl wires. Such preparations may be made with intact or detached antennae. Then the preparation is stimulated by introducing a volatile into a continuous airstream passing over the antenna (Guerrin and Visser, 1980) or the preparation can be linked to a gas-liquid chromatography (GLC) separation system allowing simultaneous recording from the GLC and an electroantennographic detector (EAD) (Arn, Stadler, and Rauscher, 1975; Guerrin and Stadler, 1982).

Where applicable, the principal technique now used for electrophysiological studies of contact chemoreceptors is tip recording from individual sensilla (Hanson, 1983; Stadler, 1984, Frazier and Hanson, 1986). Each record then represents the activity from the sensory cells contained within the individual sensillum. Tip recording utilizes a fluid-filled recording electrode that holds the stimulating solution placed over the tip and distal pore of the sensilla. Thus the electrode provides the stimulus and acts as an electrolytic bridge to an amplifier. A reference electrode is placed at a distal point of the body or preparation. A disadvantage of tip recording is that the test substance must be in aqueous phase (Stadler, 1982). An alternative, but more difficult, technique is sidewall recording in which a tungsten or glass microelectrode is inserted through the cuticle of the receptor (Hanson, 1983) allowing recordings to be made independent of the stimulus application.

The single pore of contact chemoreceptors allow for an electrical connection to be made between the sensory structure and a recording electrode without any need to penetrate the cuticle mechanically. Unfortunately, with the numerous and much finer pores of olfactory receptors (Mustaparta, 1984) this is not possible; the electrical resistance between sensillum and electrode is too high. Recordings may be obtained by electrodes penetrating the sensillum near the base and into the receptor lymph surrounding the dendrites, with a reference electrode positioned close by, often in the same antennal segment (Frazier and Hanson, 1986). Alternatively, a glass capillary electrode may be placed over the cut tip of the sensory hair (Kaissling, 1974; de Kramer, 1985). Both these techniques require ultra-micromanipulators and high-resolution optics. Stimulus delivery systems for single-cell recordings are described by Frazier and Hanson (1986) and are similar to those used in EAGs.

Most studies of Lepidopteran olfactory receptors have focused on those of larvae (Dethier, 1980a; Dethier and Crnjar, 1982; Schoonhoven, 1987) or on pheromone receptors of adult males; sensilla trichodea (O'Connell et al., 1983; Mustaparta, 1984; de Kramer, 1985; Frazier and Hanson, 1986; Meng, Wu, Wicklein, Kaissling, and Bestmann, 1989; Kaissling, Meng, and Bestmann, 1989). The latter hair sensilla are numerous and often large, making manipulation possible. Their fine structure is described in general terms by Mustaparta (1984) and for specific Lepidoptera by Steinbrecht (1980; *Bombyx mori*), Grant and O'Connel (1986; *Trichopluisia ni*) and Keil (1989; *Manduca sexta*). de Kramer (1985) provides a detailed description of the electrical circuitry of an olfactory sensillum and the techniques necessary to define it.

Single-cell recordings from olfactory sensilla have been achieved with several adult Lepidoptera (references above; Rumbo, 1981), including some noctuids: *Pseudaletia unipuncta* (Priesner, 1979; Mustaparta, 1984), *Agrotis segetum* (van der Pers and Lofstedt, 1983); and *Trichoplusia ni* (O'Connel et al., 1983; Grant and O'Connell, 1986), the latter species being particularly amenable to electrophysiological study. Although Jefferson, Rubin, Mc-Farland, and Shorey (1970) concluded that the morphology, numbers, and distribution of antennal sensilla in *T. ni* and *H. zea* were similar, the olfactory chemoreceptors on the antennae of *Heliothis* spp. have not proven amenable to the manipulation necessary for standard electrophysiological techniques (e.g., tip or sidewall recording) (E. Rumbo, personal communication). Consequently, the sensory physiology of adult *Heliothis* has not been studied extensively. The use of tungsten microelectrodes inserted at random in areas of the antennae has produced some results, but generally studies of adult antennal receptors of *Heliothis* have been unproductive. The little work done with adult receptors has concentrated on tip recordings from contact chemoreceptors (styloconic sensilla) on the proboscis (Blaney and Simmonds, 1988) and chemoreceptors on the ovipositor (M.J. Rice, unpublished). Surprisingly, there has been little attempt to use crude EAG techniques to characterize the responses of adult *Heliothis* to host-plant volatiles. This is an area that requires considerably more effort. More work has been done with larvae of *Heliothis* and other noctuids, than with adults, particularly with maxillary receptors involved in gustation (Schoonhoven, 1987). Indeed larval noctuids, particularly *Spodoptera exempta* and *S. littoralis* have been used widely in studies of the neurophysiological basis of host-plant selection and acceptance.

VI. Predator Evasion in Noctuid Moths: A Model System in Neuroethology

Neuroethologists are interested in understanding behavior in terms of the reception of environmental signals by the peripheral nervous system, and the transformation of this sensory input to motor output by the central nervous system. Therefore, they are on the lookout for animals that display behavior that is reliable and repetitive enough to be quantified, where the sensory inputs are well defined, and where the activity in the nervous system is easily recorded and simple enough to shed light on the mechanisms that generate the behavior. In this regard, the avoidance behavior of flying tympanate moths such as *Heliothis, Catocala,* and *Agrotis* in response to the echolocating calls of a predatory bat is proving to be a model system in neuroethology (Roeder, 1965).

There are several reasons why noctuid moths offer advantages for the cellular analysis of behavior. First, and most important, the behavior itself has been described in great detail. The main reason the behavior is so interesting to neuroethologists is that it is an escape behavior—the survival of the insect depends on it avoiding predatory bats, and thus nervous circuitry within the moth is adapted to organizing quick "life and death" decisions. Second, the auditory system of the moth is dedicated—generally moths do not use their ears for purposes other than listening to, and avoiding, bats. Third, the peripheral nervous input on which the behavior is based is exceptionally simple— there are only two receptors in each ear of a noctuid moth such as *Heliothis.*

1. Auditory Behavior

First described by Roeder and Treat (1957), this behavior consists of several phases that are expressed sequentially depending on how far away the bat is. When the bat is some distance away, the moth performs a "turning away" from the sound source; when the bat is closer, and the moth is in imminent danger, the moth either folds its wings together and drops to the ground ("passive drop"), or dives vigorously to the ground ("power dive"). These spectacular responses can be evoked by playing synthetic, electronically generated bat calls to the moth. By altering various parameters of the signal (intensity, repetition rate, duration) researchers have been able to determine what aspects of the echolocating signal are responsible for evoking the different phases of the moth's response. The intensity of the signal is important because it tells the moth how far away the bat is. The repetition rate and duration are important because they tell the moth what phase of the hunt the bat is in. Bats exhibit three phases in their hunt, that is, the "search" (or tracking) phase, the "approach" phase, and the aptly named "terminal" phase (Miller, 1983). Both the repetition rate and duration of sound pulses emitted by the bat are specific to each phase, and the listening moth recognizes the phase on the basis of these same parameters. The appropriate course of action must be decided in instants if the moth is to survive.

The next step in the analysis of avoidance behavior involves recording from the flight muscles that move the wings. Such recordings allow us to see how muscle activity is coordinated to produce turns, and importantly, they provide the first insight into the outputs of the central nervous system to flight muscles in response to bat calls. By inserting fine wires (25–50 μm in diameter, insulated except at the tip) into known flight muscles in a moth that is suspended above the ground in a wind stream, it is possible to record the electrical activity in each muscle as it twitches during flight (the electromyogram or EMG) (Roeder, 1951). There are sets of muscles that move the wing up (elevators), antagonistic muscles that move the wing down (depressors), and other muscles that change the pitch of the wings (forwards—pronation; backwards—supination) for steering maneuvers. By recording from a sufficient number of muscles simultaneously, it is possible to build up a picture of the pattern of activity during straight-ahead flight, and turns (Kammer, 1971), and compare this with behavioral data obtained by other means—such as measuring the aerial wake produced by the flapping wings (Roeder, 1967). Briefly, what happens in turning flight is that muscles that elevate the wing on the side from which the sound is coming are excited strongly, as are muscles that depress and change the pitch of the wing (pronation). Therefore, the wing is extended and beats with a larger amplitude. On the opposite (contralateral) side to the sound source, elevator and some depressor muscles are inhibited or receive only weak excitation, the wing pitch is changed also (supination), and the wing partially folded. The wing on this side, therefore, beats with reduced amplitude, and the result is a turn away-from the sound source (Roeder, 1967; Kammer, 1971). What is the auditory input from the ear and how is it transformed into such a motor output? In order to answer this we need access to both the peripheral and central nervous systems (CNS).

2. The Peripheral Auditory Nervous System

It is in the design of its auditory pathway that the moth offers a great advantage over other tympanate insects for neuroethological studies. In noctuid species such as *Heliothis*, *Catocala*, and *Agrotis*, the entire behavior is based on auditory input provided by only 2 receptor cells in each ear (compared with 100 in grasshoppers, and 1,500 in the cicada). This great simplicity makes it much more likely for electrophysiological and anatomical studies to explain how the moth arrives at the behavior we observe. The response properties of the two auditory receptors (A1, A2) were first described by Roeder and co-workers on the basis of extracellular recordings made with fine wire hooks supporting the tympanal (auditory) nerve (Roeder, 1965). This is still a very good method for obtaining reliable data quickly on what information the ears are providing the CNS—it can even be used in the field (Fullard, 1984). With just a pair of fine-wire, bipolar, hook electrodes supporting the tympanal nerve, isolated from body fluids with vaseline, and connected differentially to an amplifier, it is possible to obtain recordings of the two auditory afferents and the nonauditory B cell, which is probably a wing-hinge stretch receptor in atympanate moths (J. Fullard, personal communication). The three units can be distinguished easily on the basis of their spike size and this is consistent from preparation to preparation.

What can such recordings tell us about moth audition? (1) The two receptor cells are sensitive to a wide range of sound frequencies – from under 10 kHz to over 100 kHz – and this covers the range over which bats signal. However, recordings from moths in different geographical locations show that there can be variation in the range responded to, with best sensitivities restricted to the frequencies over which the bats in that particular locality signal (Fullard, 1987). (2) In all noctuid moths studied, the sensitivity of the A1 cell is always about 30 dB greater than that of the A2 cell. The physiological characteristics of the two cells are set so that when the sound level is high enough to saturate the response of the A1, the A2 just begins to respond. By dividing up the intensity range in this way (range fractionation), the two receptors are able to provide the CNS with auditory information over all the intensities at which bats signal. This has important consequences for behavior because the moth gauges the distance of a hunting bat partly from the intensity of the signal. At intensities where just the A1 cell is active, the moth begins its turn away; activity in the A2 cell occurs at intensities where we observe the power dive or passive drop (Roeder, 1965; Miller, 1983). Clearly, activity in each receptor cell spells out a specific message for the moth, and in "turning away" for example, we are witness to a complex behavior that is being driven by the activity of just the single A1 cell in each ear. (3) Moths have an ear on each side of their body, which means that they are able to compare the activity of the receptor cells on each side. The body shields the ear on the contralateral side to sound so that a signal of lower strength reaches that ear. This generates a difference in the number, and onset, of spikes in the response of the A1 receptor on each side. Behavioral observation tells us that the central nervous system is able to use these differences to resolve the direction of the incoming signal (Roeder and Payne, 1966; Roeder, 1967). (4) The receptor cells respond to the individual sound pulses in the echolocating call of the bat and record the temporal pattern of the signal – for example, the duration and repetition rate of sound pulses (Surlykke, Larsen, Michelsen, 1988). This information forms the basis of the moth's recognition of which phase of the hunt the bat is in, and, consequently, how close to capture the moth is! Thus the data provided by the simple extracellular recording method helps us understand what the ear is telling the central nervous system. Understanding behavior, however, requires that we know how neuronal circuits within the central nervous system integrate this information and send the appropriate messages out to the wing muscles. This can only be approached using intracellular recording and staining techniques.

3. The Central Auditory Nervous System

A. Techniques of Recording and Staining Nerve Cells

In order to understand how central circuits are organized we need to be able to identify neuronal elements in these circuits – both structurally and functionally. The first step in this process involves finding out where the auditory receptors from the ear end in the CNS—this is because one would expect to find postsynaptic cells in this same region. The technique first used by Surlykke and Miller (1982) to demonstrate the central projections of the A1 and A2 cells (in *Agrotis*) is called *cobalt backfilling*—it is still one of the most powerful tools in neuroanatomy. The technique involves cutting the nerve carrying the axons of the receptors and dipping the end of the nerve leading towards the CNS in a small drop of cobalt chloride (Bacon and Altman, 1977). The drop acts as a pool of high-cobalt concentration, and the metal is taken up by the axons in the cut nerve ending and transported by diffusion along the nerve, filling all branches of the axon terminals in the CNS. After some hours, the CNS is dissected carefully from the animal and the cobalt in the neurons is precipitated as cobalt sulphide, which is a black salt and, therefore, makes the neurons visible under the microscope. Subsequent histological procedures can "intensify" the image of the cell by using the cobalt salt molecules as foci for silver aggregation akin to photographic development. The staining is permanent and allows subsequent histological sectioning of the CNS. The procedure can be used to visualize any class of neuron that has an axon in either a peripheral nerve root (sensory or motor neurons) or connective of the ventral nerve cord (interneurons). By staining in the opposite direction (out into the periphery) it is possible to reveal the innervation of flight muscles by motor neurons.

The ears of noctuid moths are located on the first abdominal segment on each side of the body, and the auditory nerve joins the CNS in the metathoracic ganglion as part of peripheral nerve root IIIN1. In fact, in moths, instead of each thoracic segment having its own ganglion, the ganglia of the meso- and metathoracic segments are fused into a so-called pterothoracic ganglion. Surlykke and Miller (1982) showed that the axons of the A1 and A2 cells (and the B cell) entered the metathoracic ganglion and had distinctive branching patterns there, but, surprisingly, found that whereas the A2 (less sensitive) cell ended in the metathoracic ganglion, the A1 cell projected anteriorly probably to the brain. One conclusion from this result is that there should be centers processing auditory information in each ganglion on the way up to the brain.

Knowing where the afferents terminate gives us an indication of where we are to place our intracellular electrodes to find postsynaptic cells. Intracellular recordings enable us to examine the subthreshold electrical events occurring within a cell, and thus compare the information being received with that being sent on further to other cells, that is, integration. Although the first intracellular recordings from central auditory neurons in the moth were made many years ago (Paul, 1974), the lack of dye-filling techniques at the time meant that the recorded neurons could not be identified structurally. Since then there have been many improvements in both recording and staining techniques and the central circuitry is being unravelled. Intracellular recordings these days are made with microelectrodes made out of borosilicate glass and drawn to tip diameters of 0.1 μm on advanced pullers. The glass capillaries contain a fiber to enable the electrolyte to fill the tip completely, thus lowering electrode resistance and improving the signal-to-noise ratio of recordings. The electrode is connected to a special high-impedance direct current (DC) amplifier with a built-in bridge circuit to allow current to pass while recording the cell's activity. This current-passing facility is important because many of the new dyes are electrically charged and injected iontophoretically into a nerve cell.

The key to unravelling a neuronal circuit is to be able to record simultaneously from several neurons, and to establish which ones are connected synaptically with one another. Speaking colloquially, it is like sticking knitting needles into a bowl of spaghetti in the dark, and trying to impale two strands without damaging them, and then hoping that they are connected. This analogy is not all that far-fetched. The branches of the neurons into which we stick the electrodes are only about 5 μm in size, and if the hole we make in the cell with our electrode to record its electrical activity is too large, the cell simply "discharges" and dies. Not only that, but one has to find at least one of the "strands" again in another preparation and see what further neurons it is connected with.

Of the new dyes, the most spectacular results are obtained with such fluorescent dyes as Lucifer Yellow (Stewart, 1978). The Lucifer Yellow is dissolved in water (or lithium chloride to reduce electrode resistance) and the glass electrode is filled with the dye (by capillary action along the internal fiber), so that the dye is used as the electrolyte as well as to stain the cell. The intracellular recording is made and the dye then is injected by passing a negative current into the cell. Following fixation in formaldehyde and alcohol dehydration, the CNS is cleared of alcohol in methylsalicylate and illuminated by ultraviolet (UV) light under a fluorescence microscope – the cell containing Lucifer Yellow fluoresces a brilliant yellow (500 nm wavelength) against the dark-green background of the ganglion. Regrettably, this fluorescence is not permanent in wholemount tissue and fading occurs rapidly on exposure to UV radiation (even in normal room lighting), so the stained cell must be photographed immediately, and then embedded in araldite for sectioning – a procedure that stabilizes the fading.

B. The Physiology of Auditory Neurons in the Central Nervous System

What do we know of the central circuitry mediating avoidance behavior as a result of these new methods? Boyan and Fullard (1986,1988) and Boyan, Williams, and Fullard (1990) have identified about 15 auditory interneurons located in the pterothoracic complex. By "identified" we mean that the individual neuron was recorded and stained in many preparations, and always looked the same, and responded to sound in the same way in every preparation. The first thing we were struck by is that no matter what species of noctuid moth we worked with (*Heliothis*, *Catocala*, *Agrotis*), the same neurons (structurally and functionally) were present in all. So at the moment, it looks like the auditory pathways of all noctuid moths are built along the same lines. The auditory interneurons we identified have branches in the same region of the ganglion as the A1 and A2 auditory receptors (Boyan and Fullard, 1986, 1988). By recording simultaneously, for example, from the A1 afferent and successive interneurons, we found some interneurons that received their information directly (monosynaptically) from the A1 receptor (Boyan and Fullard, 1988). We then asked what happened to the information that crossed the synapse, and the example we give below shows how one can relate physiological data from intracellular recordings to the auditory behavior of the moth.

One very important problem the moth must solve is to distinguish the echolocating cries of a bat from the cacophony of sounds that emanate from the nighttime environment — mainly from other insects calling at ultrasonic frequencies. When we compared the response of the A1 afferent and an interneuron to a range of sound intensities, we found that subthreshold activity in the interneuron did not begin to summate until the spike rate in the receptor reached a certain level. This level was quite high because the events evoked in the interneuron by each afferent spike were extremely short. The synapse was functioning to prevent activity in the afferent below a certain rate from being passed on to higher circuits — it was filtering the input. Then in order to bring the interneuron itself to spiking, the afferent had to fire at a much higher level (spike rate) — one that would result from a sound pulse with a sharp onset — like that emitted by a bat. In fact, Roeder and Payne (1966) had predicted the level at which summation should occur based on behavioral reactions and afferent recordings, and it is very close to that recorded by us in the synapse we investigated. So the synapse appears to be adapted to filter out low-level signals from the environment that do not produce a critical spike rate in the A1 afferent, and along with it, the requisite level of summation in a central neuron. We interpret this as meaning that the synapse is part of the circuitry telling the moth what constitutes a "bat" and "non-bat".

Now that we have begun identifying interneurons in the CNS, we can say that the noctuid moth appears to be unique among tympanate insects because of the organization of its auditory pathway. There are only two receptors in the ear, yet many auditory interneurons in the CNS — an extreme case of divergent information flow. How is information flow in the auditory pathway organized, and how does it relate to avoidance behavior? Most of the interneurons we have recorded ascend from the metathoracic ganglion to the brain — in parallel with the A1 receptor. As the receptor, some of the interneurons respond tonically to an auditory stimulus and so can encode the temporal duration of a sound pulse. Such neurons were recorded extra-cellularly previously and termed *repeater cells* (Roeder, 1966). Other inter-neurons respond phasically to a stimulus — that is, they fire only once with the onset of a stimulus ("pulse coder"), and so can encode only the rate at which stimuli occur. Our intracellular recordings are providing evidence that two interneurons may respond differently to a sound stimulus (tonically or phasically) despite both receiving their input directly from the same A1 afferent (Boyan et al., 1990). This means that the differences in their re-sponses are due to the way the interneurons themselves integrate afferent information, and, hence, to the genetically determined membrane character-istics of these cells. The moth interneurons, therefore, can encode all the elements (intensity, pulse duration, pulse repetition rate) that behavioral experiments have shown are necessary to recognize a bat signal, gauge the distance to the bat, and ascertain the phase of the hunt. In recording the electrical properties of these cells, and the way they interact with others in the circuit, we are seeing the evolutionary adaptation of the moth senso-ry system to the selection pressure exerted by the echolocating calls of the bat.

C. Outputs to Motor Pathways for Flight

The final step in avoidance behavior involves the signal being sent to the flight muscles by motor neurons (Madsen and Miller, 1987). Currently, we are looking at the information motor neurons receive from the different types of interneurons described above (Boyan et al., 1990). The forewings are the primary power-generating structures during flight, and the motor neurons that innervate the muscles driving the forewings are located in the mesothoracic ganglion (Kondoh and Obara, 1982; Rind, 1983). We have identified a set of auditory interneurons in the mesothoracic ganglion that have branches in the same regions of the ganglion as do these motor neurons (Boyan et al., 1990). One interesting feature of the interneuron responses is that they are all phasic to sound stimuli, whereas the population of inter-neurons located posteriorly in the metathoracic ganglion contains tonic as well as phasic types. It appears that the closer we get to the motor pathway, the more phasic the responses of interneurons. Remember that only the single A1 afferent ascends the ventral nerve cord to the brain from each ear, and in so doing it provides these different populations of interneurons with the same primary auditory input. Phasic responses may be interpreted as instantaneous "pulses" that can be used to trigger events — for example the decision to turn away or fold the wings and drop. Although still very specu-lative, our results suggest that the two groups of interneurons, though in neighboring parts of the CNS, may contribute to different steps in the overall expression of avoidance behavior.

VII. Conclusion

The stereotyped evasive response of a flying moth to the acoustic signal of an echolocating bat is an excellent example of how behavior can be understood in terms of the activity of the nervous system. Even in this system, the simplicity evident in the peripheral auditory pathway gives way to quite complex integration in the central nervous system. Nevertheless, we have been able to relate the properties of both a single synapse, and the overall organization of information flow in the pathway as a whole, to aspects of avoidance behavior. It is important to note that the level of understanding we have reached to date comes from combining anatomical, physiological, and behavioral studies. Similar approaches will be needed to elucidate the more complicated chemoreception of host plants and mates, but is essential if we are to understand oviposition and mating behavior.

Our studies have brought a further interesting by-product. When we looked at the neuroarchitecture of the moth CNS, that is, the pathways taken in the CNS by nerve fibers originating from all over the insect, we found that the internal structure of the moth CNS is very similar to that found in such hemimetabolous insects as the cricket, grasshopper, or praying mantis (Boyan et al., 1990). In addition, we found that the auditory pathway occupies the same relative position in the CNS in all these atympanate insects (Boyan, 1989). There appears to have been a particular design to mechanosensory pathways of the ancestral insects of these groups that has been retained for auditory neurons in tympanate insects even though the ears may be located on different parts of the body (forelegs, thoracic midline, abdomen). Thus analysis of the structure and function of the auditory pathway in noctuid moths is providing insights into the evolution of audition in the insects as a whole.

Acknowledgments. We are grateful to Lindsay Barton-Browne, Roger Traynier, Eric Rumbo, and R.M. Henning for valuable criticism of earlier versions of the manuscript. GPF is especially grateful to LBB for valuable advice and guidance in the area of insect behavior over several years. We also thank Carol Murray (CSIRO, Black Mt. Library) for her assistance in obtaining reference material.

Chapter 12

Measuring Insecticide Resistance

R. V. Gunning

I. What is Insecticide Resistance?

When the failure of an insecticide (applied at its normal rate) to control a population of insects is due to a genetically transmitted capacity to tolerate more insecticide than normal, then insecticide resistance is said to have occurred. In field resistance, these simple biological factors are complicated by such economic considerations, as what level of pest presence is deemed acceptable.

II. Why Measure Resistance in *Heliothis*?

Heliothis armigera and *H. punctigera* are the most important pests of field crops in Australia. Insecticides are used primarily for their control, especially on cotton. While there has been no difficulty controlling *H. punctigera*, *H. armigera* has a long history of insecticide resistance in Australia, first to DDT, in the early 1970s (Wilson, 1974) and more recently to the pyrethroid insecticides (Gunning et al., 1984). Resistant *H. armigera* have posed a severe threat to the economic production of cotton in Australia always. Due to the high cost of insect control, insecticides already comprise one third of total production expenditure.

Resistance monitoring is an essential part of any *Heliothis* pest management program. To diagnose resistance, insecticide testing of a suspected resistant strain needs to be compared with results from strains known to be susceptible. However, there is considerable variation in response to insecticides even in susceptible populations. Therefore, extensive testing of these populations is needed to establish with reliability the natural range of response to insecticides. This baseline data can be obtained only by a long-term commitment to routine monitoring and testing programs. As was evident in Australia in 1983, baseline data can enable speedy diagnosis of resistance. As a result of an established national *H. armigera* resistance monitoring program, the accumulation of reference data enabled a swift detection of pyrethroid resistance and the adoption of a resistance management strategy within 6 months of the confirmation of resistance.

The goals of resistance monitoring in *Heliothis* are to measure insecticide susceptibility levels and to anticipate field control problems and field failures. Bioassay data can give indications of heterogeneity of response in a *Heliothis* population toward an insecticide, long before field selection of those resistant individuals cause control failures.

III. Problems of Resistance Monitoring

The generation of reliable baseline data for *Heliothis* can present problems because insect populations must be collected widely to take into account variables such as geographic location, host plant, season, and so forth. Reliable resistance measurement requires the collection of sample sizes large enough so that resistant individuals, at a low frequency can be detected. Obviously, the smaller the sample size, the more difficult this is to achieve. Programs aiming to detect resistance prior to control failures require great precision in estimating the frequency of resistant individuals and with the adoption of resistance management strategies, similar precision is required to detect small changes in resistance frequency. Roush and Miller (1986) have suggested that a resistance monitoring program should be designed to detect resistant individuals at a 1% frequency. Presently, we do not understand statistical considerations that affect the accuracy of monitoring, especially of low-level resistance situations, when numbers to establish reliable results are very large.

IV. Techniques for Measuring Resistance

For general information about insecticide bioassay methods and statistical analysis of results, Busvine (1971) and Finney (1971) are essential reading.

Bioassays play a vital role in detecting resistance and in evaluating changes in resistance frequencies of field populations. Such bioassays must be a means of reliably, quickly, and cheaply distinguishing between resistant and susceptible individuals. Bioassay techniques for *Heliothis* vary according to the type of insecticides and the life stage of the insect at which it is directed. Larvae are usually the target with contact insecticides and most bioassays reflect this, however ovicides must be tested against eggs, stomach poisons delivered via food, and so forth.

The results of a bioassay (the mortality recorded over a range of insecticide concentrations) are plotted as dosage mortality curves and analyzed by probit analysis (Finney, 1971).

The most common measure in toxicology (and statistically the most reliable) is the dose at which 50% are killed (known as the LD50, Fig. 12.1). Insecticide LD50s are widely quoted as an indication of their effectiveness. The slope of the straight line is also an important statistic, because it is a measure of variance in the population. The shallower the slope, the greater is the variation in susceptibility to the tested insecticide. Resistance measurement for different insect strains generally involves a comparison of LD50s and slopes, but the frequency of resistant individuals must be high before the LD50 is changed appreciably (Fig. 12.1). The slope and the LD95 are better indicators of heterogeneity. These limitations were recognized at the onset of pyrethroid resistance in *H. armigera* in 1983 when a single discriminating dose technique was adopted by (Gunning et al., 1984). The dose, equivalent to the LD99.9 of susceptibles, has been a very useful and an efficient way to monitor the subsequent spread of pyrethroid resistance (Gunning and Easton, 1989).

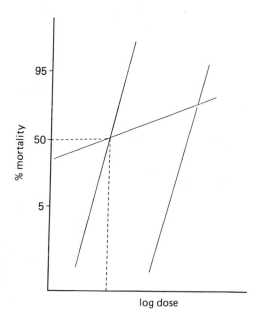

Figure 12.1. Dosage mortality lines showing the relationship between changes in LD50 and slope with resistance frequency in populations. (See text for details.)

V. *Heliothis* Bioassay Methods

1. Contact Insecticides

A. *Third Instar Testing*

Most *H. armigera* resistance data has been obtained by a standard test method; the topical application of contact insecticides on third instar larvae in a procedure similar to that recommended by the Entomological Society of America (Anon, 1970), using acetone solutions of technical grade insecticides. The dosage mortality curve requires at least 150 larvae per chemical. These large numbers can be obtained only by laboratory breeding of field-collected insects. However, the single discriminating dose technique, requires less insects (approx. 50) and can be used to screen field-collected *Heliothis* directly.

B. *First Instar and Neonate Testing*

A recent development in testing contact insecticides has been to use them against first instar larvae (R.V. Gunning, unpublished). Eggs, just prior to hatch are attached to cotton leaves, then dipped or sprayed with aqueous insecticide concentrations and held at constant temperature and high humidity. On hatching, the neonate larvae receive a dose by contact with the residue on the leaf. The technique can be used to distinguish pyrethroid resistant and susceptible *H. armigera* on the basis of the LD50. A discriminating dose for resistance, the LD99.9, can be adopted also.

A first instar bioassay has advantages over the standard third instar method. First, less *Heliothis* rearing saves time and money. Second, it is possible to test many more individuals per concentration which is important when resistance is at a low frequency. On the other hand, it cannot be used to screen field individuals directly unless collected from such *H. armigera* specific crops as maize. (At present nonlethal differentiation between *Heliothis* spp. eggs is not possible.)

C. *Adult Testing*

Adult *Heliothis* can be tested for insecticide resistance also, although the moth scales present some problems for insecticide contact. Two methods have been tried in Australia: (1) topical application of insecticide to the compound eyes (N.W. Forrester, personal communication) and (2) adult vial tests (J.C. Daly, personal communication). The adult vial test is based on that described by Plapp (1979). Glass scintillation vials are coated evenly with acetone/insecticide solutions. Pheromone-trapped male moths are placed individually into treated vials and allowed to come in contact with the treated surface at constant temperature and humidity. Mortality is assessed after 24 h.

2. Ovicides

Some insecticides, such as chlordimeform and methomyl, are known to kill *Heliothis* eggs. The techniques (R.V. Gunning, unpublished) that have been used for egg bioassay are similar to those described above for first instar testing. Host-plant leaves, with fertile eggs attached (as determined by the presence of a head capsule), are sprayed or dipped with formulated ovicides. The eggs are allowed to develop then at constant temperature and humidity. Egg mortality is assessed with the aid of a binocular microscope.

3. Insecticides with Novel Modes of Action

Such noncontact insecticides as stomach poisons (e.g., thiodicarb, insect growth regulators, etc.) require that the insecticide be fed to the *Heliothis* larvae. Host-plant leaves (e.g., cotton) are sprayed using a Potter Tower with the formulated insecticide concentrations and standard-size leaf discs are fed to single third instar larvae. Larvae usually take some time to die. If necessary, the precise dose ingested can be calculated from the remaining leaf material.

4. Which Methods Discriminate Best Between
Resistant and Susceptible *Heliothis*?

McCaffery et al. (1988) who examined discrimination between pyrethroid-resistant and susceptible *H. armigera*, found that the adult vial test did not provide adequate discrimination, whereas topical application and leaf residue bioassays gave consistently better results. Leaf residue and topical application methods are best for detecting the presence of resistant individuals accurately, but they are also the most labor-intensive. Despite poor discrimination, the adult vial test is inexpensive and straightforward and for that reason has been used extensively in the United States. It is known that adult insecticide susceptibility varies with moth age, even in resistant individuals (N.W. Forrester, personal communication). Because light and pheromone traps collect moths of uncertain age, the adult vial test is not recommended.

VI. Relevance of Bioassays to the Field

Resistance is usually measured in the laboratory. At present, there are no suitable methods that can test *Heliothis* for insecticide resistance in the field directly (see Chapter 13). Problems of sampling, insect size, accurate dos-

age, and temperature-dependent toxicity make field bioassay methods difficult to develop.

Unfortunately, the level of resistance shown by a bioassay does not always
reflect field-control problems. Field insecticide rates are usually "robust"
and kill all, but the larger resistant individuals. This was illustrated in *H.
armigera* by Daly et al. (1988) who showed that the standard rate of fenva-
lerate was only selective to larvae more than three days old. The method of
determining the critical resistance frequency where field control is impaired
is presently unknown, but very important.

VII. Current Resistance Status of *Heliothis* Species in Australia

H. armigera has a long history of resistance in Australia: to DDT in the early
1970s (Wilson, 1974) and to the pyrethroids in 1983 (Gunning et al.,1984).
In 1989, pyrethroid-resistance frequencies in both sprayed and unsprayed
populations were very similar (Forrester, 1989; Gunning and Easton, 1989).
From 1985 to the present, approximately 40–50% of field individuals tested
have been found to be resistant. Resistance also seems fixed in unsprayed
populations.

Increased use of endosulfan has led to a resurgence of resistance to this
insecticide in *H. armigera*. In 1989, a low (15–30%) level of resistance to
endosulfan was common throughout eastern Australia. Carbamate resistance (methomyl, carbaryl) is similarly widespread (R.V. Gunning, unpublished). These resistance frequencies have not resulted in control problems
yet. *H. armigera* have incipient resistance to some organophosphates.

H. punctigera are susceptible to all insecticides (R.V. Gunning, unpublished).

Chapter 13

Methods for Studying the Genetics of Populations of *Heliothis*

J.C. Daly

I. Introduction

This chapter outlines the different methods for studying population genetics of *Heliothis*. It is assumed that the reader is an entomologist with a background in ecology and a limited exposure to the concepts and techniques of population genetics. The first part of the chapter discusses some of the concepts that are necessary to master and some of the unresolved issues that must be understood before the theory can be applied to real populations. The second part considers the kinds of questions that population genetics studies can address. Finally some of the methods and their pitfalls are then described.

II. What Is Population Genetics and Why Study It?

Population genetics is the study of changes in, or stability of, the frequencies of genes in populations, and the rate of divergence of gene frequencies in populations wholly or partly isolated from each other. As such, it is part of the discipline of evolutionary biology. Population genetics theory remains an abstract exercise unless the frequency of alternative alleles at various loci can be determined in different populations and at different times. For some kinds of variation this is not always possible, for example, fecundity or certain color polymorphisms, yet this genetic variation is often of more interest than are single gene characters. This chapter deals with a slightly larger topic, that of the genetics of populations. In this way, we may use the theory and methodologies of population genetics to address problems more correctly grouped under systematics, quantitative genetics, and ecology.

Population genetics can contribute to an understanding of the biology of *Heliothis* for two reasons. First, many important phenomena arise as a consequence of genetic differences between individuals, for example, insecticide resistance and host races. Differences between individuals in adaptive characters, such as the tendency to diapause or to disperse, may also have a strong genetic basis. Second, aspects of the biology of an organism, such as its mobility, spatial patterns of oviposition, deviations from random mating, and so on, can lead to an accumulation of genetic differences between populations. Inferences about the structure of a population can be made, with care, from a study of its genetic structure, defined here as the spatial pattern of gene frequencies (see below).

To use population genetics theory to gain an insight into the biology of *Heliothis*, it is crucial to understand what a geneticist means by terms such as *migration*, *gene flow*, *population*, and *population size*. A Mendelian population is an interbreeding group of organisms and constitutes a gene pool. In the simplest genetic models, a population is infinitely large with random mating between individuals and an equal sex ratio. All adults have an equal probability of leaving offspring, and the generations are nonoverlapping. Real populations deviate from this ideal model; in general, these deviations increase the rate at which changes occur in gene frequencies. Such changes are said to be a result of genetic drift.

To the population geneticist, a population is more than a collection of individuals that can be counted. It is defined by a number of parameters: the size of the breeding population and fluctuations in this number over a number of generations, the sex ratio, the variance in the number of progeny of breeding individuals, deviations from random mating within the population, the degree to which the generations overlap, and the proportion of reproductively successful individuals that immigrate each generation; that is, by those parameters that have an effect on gene frequencies in the population.

The combined effect of these parameters is summarized in the concept of effective population size, N_e. This is an abstract number that adjusts the actual number of individuals in a population to take into account the deviations from the ideal population. It is this number, N_e not the actual population size, N, that is used in population genetics theory (although equations often use N and N_e interchangeably). Formulae for calculating N_e are found in Crow and Kimura (1970) and Cavalli-Sforza and Bodmer (1971). Although the theory of population genetics is well described mathematically, it has been difficult to apply and test the theory in natural populations because many of the parameters in these equations are not easy to measure.

Often it is not possible to determine exactly where one population ends and another begins, nor can the limits to a population be drawn around defined areas. The indeterminate nature of a population need not be a problem because any unit of population is defined both by its deviations from the ideal population and by the amount of immigration into the population. The magnitude of each of these will vary with the unit of population chosen for study. The smaller the unit, then the greater the amount of migration between populations will be.

If you are interested in changes in gene frequency over a single generation, then the unit of population may differ depending on the time of year. Consider the cotton-growing area of the Namoi Valley, New South Wales in Australia. In spring, when *H. armigera* emerges from diapause and colonizes new crops, the unit of population may be an entire cropping area or large parts of it. During the period of peak flowering, when *H. armigera* is considered to be more sedentary (Wardhaugh et al., 1980), the unit of population may be a field or a group of adjacent fields.

Gene flow (= migration in genetics) occurs when individuals disperse between populations and successfully reproduce in their new locality; that is, they contribute to the gene pool. Although gene flow is a result of dispersal, its relative importance in producing genetic change in a population is not correlated necessarily to the magnitude of dispersal between populations because genetic homogeneity between populations can be maintained with very little gene flow under some conditions (Speith, 1974; Allendorf & Phelps, 1981). Adequate information about gene flow can be made only from detailed observations of both dispersal and the genetic structure of populations (Slatkin, 1981,1987).

Gene frequencies in a given population can change as a result of genetic drift, natural selection, or gene flow. All three may act together on each gene in a population. The amount of genetic drift is dependent on N_e (see above) and the amount of gene flow; it should be uniform across all loci. The magnitude of selection, however, will vary between loci. If selection is very weak, then the patterns of genetic variation will reflect the structure of the population and the genetic variation is said to be neutral. If selection exceeds the combined effects of drift and gene flow, then the variation is adaptive. It is difficult to demonstrate that variation is neutral or adaptive (Lewontin, 1974; Ewens, 1977). One exception is with some variants of hemoglobin in humans. Individuals that are heterozygous for an abnormal form of this protein are slightly more resistant to malaria than are those homozygous for the normal protein (Cavalli-Sforza and Bodmer, 1971). Generally, the variation detected by electrophoresis is regarded as neutral, whereas genes conferring resistance to insecticides are adaptive in the presence of the insecticide, but may be neutral or maladaptive in its absence.

A study of the relationship between genetic structure and population structure is part of the discipline of ecological genetics. There is not a direct relationship between a given population structure and the spatial patterns of gene frequencies because genetic drift is a random or stochastic process, unlike selection whose effect is predictive or deterministic. Thus a given genetic structure, for example, homogeneity of gene frequencies in populations over a wide geographic area, can be generated by a variety of contrasting population structures and selection regimes. This is a major problem when applying the theory of population genetics to a study of population structure. In their study of genetic variation in *Heliothis* spp. in Australia, Daly and Gregg (1985) found that gene frequencies were homogeneous over eastern Australia. By itself, this is insufficient evidence of gene flow between populations because the same pattern could also have arisen in a species in which gene flow was low between populations, but where N_e was very large (Daly, 1989).

Sometimes independent data about the biology of the species can help exclude alternative hypotheses. For instance, the presence of genes for resistance in areas in which insecticides have not been used suggests that there must be occasional long-distance migration. The data may be examined using more discriminating genetic tests. Daly and Gregg (1985) were able to infer that gene flow must be occurring between populations of *Heliothis* spp. because of the distribution of low-frequency alleles (Slatkin, 1981).

Care must be exercized when applying population genetics to a study of population structure. It can be very useful, but it can also be misleading. There is not a simple list of "do's and don'ts" as the subtleties of interpretation are too great and each study must be considered independently. Before embarking on such a study, a population geneticist should examine your program, and after you have analysed the data let him or her look at your interpretation. This is particularly true if you are examining either a genetic phenomenon such as fitness differences between different genotypes, or determining the genetic basis of some character.

III. Questions That May Be Examined

Population genetics can contribute directly or indirectly to our understanding of the following areas of biology.

1. A Study of Genetic Phenomena

Variations between organisms in such characters as morphology, or in response to such chemicals in the environment as insecticides can arise because of genetic differences between individuals. A population genetics study could measure the changes in the frequencies of genes that either determine, or are correlated with, the character or response, and the causes of these changes. This is important for such characters as insecticide resistance. The common method for observing the occurrence of resistance in a population is to measure a change in the slope and LD_{50} of a probit line (see Chapter 12). This is only a crude method of measuring gene frequency changes. If the genetic basis of the character is simple, (as it usually is in resistance), then a population genetics approach is appropriate (Roush and Daly, 1990). This may not be possible if the character is determined by many genes, or if the differences between the genotypes are small.

You may be interested in the differences between resistant and susceptible individuals for such life-history traits, as fecundity and longevity. It is not adequate to compare these traits in individuals trapped directly from the field if you are interested in a causal effect of the resistance allele on these characters. Life-history traits are likely to be affected by many different genes. Hence, to examine the effect of the genes for resistance, the background must be made homogeneous for other genes (Roush and McKenzie, 1987). This is accomplished by standard laboratory crosses over many generations in which resistant individuals are backcrossed repeatedly to an inbred susceptible line. McKenzie, Whitten, and Adena (1982) illustrate this technique. Alternatively, isofemale lines can be established (see below). However, if there are only small differences between wild-caught resistant and susceptible individuals for the trait under consideration, then it may be possible to discriminate between some hypotheses without first isolating the resistance genes into a known genetic background.

2. Identification of Sibling Species

Once populations or species become isolated from one another, genetic differences between them begin to accumulate. Species can be distinguished readily from one another at the level of the protein (Avise, 1974), the chromosome (Jackson, 1971), or DNA (Solignac and Monnerot, 1986) using techniques outlined below. Electrophoresis is a relatively simple technique for distinguishing different species because these are likely to be fixed for alternative alleles at a number of enzyme loci. It has been used to determine the species of individual eggs and larvae of *Heliothis* (Daly and Gregg, 1985). Not all lepidopteran species can be differentiated using electrophoresis (Harrison and Vawter, 1977; Brittnacher, Sims, and Ayala, 1978).

It is disheartening to find that what you thought was a single species, is actually a species complex. For example, Pashley, Johnson, and Sparks (1985) have detected two biotypes (see below) of the fall armyworm, *Spodoptera frugiperda*. The one on rice and bermuda grass appears to be genetically different from the biotype found on the more common crop host, maize. Initial breeding data in the laboratory suggest that the two strains are isolated reproductively (Pashley, 1986) and so probably are sibling species. Another example is the "races" (see below) of the fruit fly, *Rhagoletis pomonella*, that are also three valid species (Bush, 1966).

Daly and Gregg's (1985) study of *H. armigera* and *H. punctigera* in

Australia suggested this is unlikely to be the case with either of these pest species, but their survey of *Heliothis* populations was not exhaustive. The taxonomy of the Australian Heliothini is not well known so it is possible that unrecognized species are still to be described, particularly from localities in which agriculture is very recent, such as in the tropics. Also, we do not know if *H. armigera* is a single species throughout the world or a series of sibling species. Similarly, there has not been a thorough survey of either *H. armigera* or *H. punctigera* from native hosts in Australia (see Chapter 1).

Problems with taxonomy may be greater if you are interested in a study of the parasites or predators of *Heliothis*, particularly if they have not been subject to a recent taxonomic revision.

3. Identification of Biotypes or Races

Heliothis is a polyphagous insect. You may wish to investigate whether either pest species is subdivided into genetically identifiable taxa (subspecies or race) or whether individuals differ in their host preference and if this difference is determined genetically. Population genetics can be used to address some of these questions.

Terms such as *race*, *biotype*, and *subspecies* are not easy to define precisely (Berlocher, 1979). A useful distinction is that a subspecies is a broad geographical subdivision of a species that shows consistent morphological or biological differences from other parts of the species range (Mayr, 1969). The term *subspecies* is sometimes limited to those cases where there is a degree of reproductive isolation between the subdivisions, whereas a *race* refers to those subdivisions in which there is no barrier to reproduction, except geographic isolation. For example, the grasshopper, *Caledia captiva*, is uniform in its morphology throughout its distribution. Studies of chromosomal variation revealed the presence of two sibling species. One of the species could be subdivided into three chromosomal taxa. One taxa (a subspecies) was partially isolated reproductively from the other two races (Shaw, 1976). Electrophoresis could also differentiate the different taxa (Daly, Wilkinson, and Shaw, 1981).

A host race is a population that shows a preference for a specific host plant that differs from other populations (Bush, 1969). The heritability of host preference is discussed more appropriately by the discipline of quantitative genetics (Mather and Jinks, 1977). Pashley (1986) illustrates one technique for such a study. The term *biotype* is a more general one than host race and distinguishes between populations of a species that differ from one another in attributes other than just host preference, for example, in insecticide resistance or life-history traits. Gonzalez et al. (1979) discuss the concept of the biotype in pest management and biological control. Electrophoresis has been useful at detecting biotypes in a number of insect species. The best examples are in the aphids (Adams and van Emden, 1972;

Eastop, 1973) and in fruit flies (Bateman, 1976). Many so-called biotypes, however, have been shown to be sibling species (see above). Futuyma and Peterson (1985) review the use of genetic variation to study the use of resources by insects.

Biotypes and races may be detected using electrophoresis, if consistent differences in gene frequencies can be observed for a number of loci between samples collected from different localities. Caution must be exercized to exclude other factors as the cause of the genetic differences, such as a recent population crash. Also, it is important to determine if morphological or life-history traits are due to underlying genetic differences between the biotypes, subspecies, or host races, or simply represent different phenotypic expression induced by the environment. If the latter is the case, then population genetics will have little to contribute to an interpretation of the variation. If biotypes are not different genetically, then you will have to use other techniques to distinguish them (for example, pheromone composition, mating behavior; see Chapter 1).

4. Population Structure

Individuals in a population can differ from one another at many loci. Allozymes (see below) can act as markers in a similar way to that employed in a mark-recapture experiment except, rather than the mark being applied externally by the experimenter, most genetic markers are internal and only can be visualized biochemically. Genetic markers can be used to examine some ecological problems. For example, genetic studies can reveal the ways in which a species is subdivided, not necessarily into races or biotypes but into populations that are isolated relatively from one another. It can reveal this subdivision even down to a fine scale, such as a field (Daly, 1989), depending on the question being asked and the time period being considered. Richardson, Baverstock, and Adams (1986) discuss how to design a study that examines population subdivision.

If little is known about a species, then genetic studies have a number of advantages. First, the insect may be too small at some or all of its life stages to mark using standard techniques or there may be a very low chance of recapturing marked individuals. In this case, electrophoretic variants can be used to look at dispersal and gene flow (Daly, 1989; Slatkin, 1985). Second, a genetic approach can be performed quickly and relatively cheaply. Thus it may be able to provide alternative hypotheses to be tested in a longer-term ecological study.

Genetic studies cannot be done in the absence of independently derived data about the species because, as noted above, there are usually a number of explanations for a given genetic structure. Often additional information about the species can exclude a number of alternatives. For example, we know from visual observations that *Heliothis* moths do disperse between fields and that females mate over a number of nights, most likely with different males. Given these observations, microspatial variation in gene frequencies can be used to gain insight into oviposition behavior of individual females or the distances that adults disperse each generation.

Genetic markers can be used also to study aspects of the reproductive biology of organisms, such as the pattern of sperm precedence in a female that has been mated more than once. It would be necessary to select crosses in which males differed from one or another at one or more genetic markers before the test can be applied. Enzyme markers have been useful to study precopulatory isolating mechanisms in mosquitoes (Coluzzi and Bullini, 1971) and to elucidate mate identification in a colonial marine invertebrate (Grosberg and Quinn, 1986).

Traditionally, genetic studies of population structure have used such markers as chromosomal or electrophoretic variants. Variation in DNA base sequences, detectable by analysis of restriction fragment length polymorphisms (RFLPs), is another source of variation. Electrophoresis can detect mutations that only produce a difference in the total charge of the protein, whereas analysis of RFLPs can detect a much larger number of mutations (Quinn and White, 1987). This increases the amount of genetic variation that you can utilize in a study. Mitochondrial DNA (mt-DNA), may be useful also (for example, Lansman, Shade, Shapira, and Avise, 1981; Harrison, Rand, and Wheeler, 1987), and because it is maternally inherited, all individuals descended from a single female have identical sequences for this DNA, except for de novo mutations; such family lineages appear to be clones with respect to their mt-DNA. This is not so for nuclear DNA that is inherited from both parents and undergoes recombination and segregation.

Bartlett and Raulston (1982) have reviewed the usefulness of genetic markers for population studies in *Heliothis*. A number of electrophoretic studies have been undertaken. These are Sell, Whitt, and Luckmann (1975); Sluss, Sluss, Graham, and Dubois (1978); Sluss and Graham (1979); Wolfenbarger et al. (1973) and Daly and Gregg (1985). Sluss and Graham (1979) suggest that a low measure of heterozygosity in one of their samples is probably a result of immigration from a number of source populations. This is true in the adult immigrants but in the larvae, which is the lifestage that they were measuring, we would expect an excess of heterozygotes.

Examples of genetic studies of population structure in other insects include Archie, Simon, and Wartenberg (1985) and Sturgeon and Mitton (1986). General references that give the theoretical basis to evolution and population structure are Dobzhansky et al. (1977); Wright (1978); Futuyma (1979); and Lewontin (1985).

5. Genetic Techniques in Pest Management

The release of genetically altered insects to suppress populations of pest species is another application of genetics. It is not strictly part of population genetics, but it is an example of how techniques developed within one discipline can be applied in a novel way to another discipline. Various approaches have been taken and are reviewed in Pal and Whitten (1974) and Whitten (1979). The released individuals can be less fit than endemic individuals if they carry a genetic load because of chromosomal translocations (Foster, Whitten, Prout, and Gill, 1972) or are sterile, as in the case of screwworm fly (Knipling, 1979). These releases are directed at suppressing the pest population. Alternatively, the released individuals may be more fit than endemic individuals as is the case for releases of spider mites that are resistant to insecticides (for example, Roush and Hoy, 1981). The aim of these releases is to ensure that adequate numbers of predatory mites survive insecticide applications so that they can suppress populations of phytophagous mites.

Attempts have been made with *H. virescens* to produce sterile males for field release by backcrosses of this species to *H. subflexa* (Laster, 1972). The daughters of the backcross progeny carry genetic information that produces sterile males when crossed to a *H. virescens*. A pilot project on the island of St. Croix indicated that massive releases of backcross individuals should suppress budworm populations (Proshold, 1983). The feasibility of such an approach for pest control of Australian *Heliothis* is unknown.

6. Other Questions

It is difficult to summarize the wealth of questions that can be addressed using genetics, for example, Pashley and Proverbs (1981) use electrophoretic markers to monitor the quality control of codling moths in a laboratory colony. Needless to say, it is necessary that the species has some genetic variation in a convenient marker and that reasonable sample sizes be possible. Detailed genetic studies are limited at this stage in *Heliothis* because of the absence of work on cytology and the molecular biology of the organism, and because of the lack of genetic markers and linkage maps both of which are standard tools of the population geneticist.

IV. Methodologies

1. Electrophoresis

Electrophoresis is a biochemical technique that is capable of separating and visualizing allozymes and isozymes. Isozymes are enzymes that can act on the same substrate, but are coded for by different loci (genes). Allozymes are enzymes that are coded for by genetic variants, alleles, of the same locus, and usually differ from each other in their overall electric charge. With this technique it is possible to obtain allele frequencies from a number of loci, each of which is assumed to be effectively neutral. It is this variation that is often studied in population genetics.

The enzymes are separated in a supporting medium (starch, cellulose acetate or polyacrylamide) in an electric field. Details of these techniques can be found in Harris and Hopkinson (1976) for starch and acrylamide and in Richardson et al. (1986) for cellogel. Both these books are very comprehensive and cover all stages of laboratory procedure. Richardson et al. (1986) also give a detailed account of project planning, statistical methods, sample sizes, strategies, how to collect and store the data, and how to interpret the gels and analyze the results.

Electrophoresis can be performed in any standard laboratory that is moderately well equipped. You will need a power pack (up to 500 V), electrophoresis tanks (these can be made much more cheaply than the commercially available ones; see above references), a small centrifuge, a deep freeze, a refrigerator, a micropipette, a sensitive balance and such small items as plastic lunch boxes and standard laboratory glassware. For the collection of specimens you will require a liquid nitrogen tank (about 10 to 20 l) and cryogenic tubes.

Be warned that some of the components used to visualize the allozymes on the gels are carcinogenic. Seek advice on their proper storage and handling. Samples should be stored in cryogenic tubes when placed in liquid nitrogen. Alternative tubes can break in storage or explode when they are transferred.

Daly and Gregg (1985) list isozymes and buffers suitable for cellogel electrophoresis of Australian *Heliothis* spp. Techniques may need to be modified under some conditions because not all isozymes stain with equal intensity at all life stages, for example, 6-phosphogluconate dehydrogenase (E.C. 1.1.1.44) works well in larvae but can stain weakly in eggs and adults, and guanine deaminase (E.C. 3.5.4.3) bands are scorable only in eggs and larvae. Some isozymes work only if samples have not been frozen, for example, malate dehydrogenase (E.C. 1.1.1.37) and malic enzyme (E.C. 1.1.1.40), whereas other isozymes work best after the samples have been frozen. Better results may be achieved with new buffers, for example, 0.02 M phosphate pH 7. 0 buffer (Na_2HPO_4, 8. 4 mM NaH_2PO_4; Richardson et al., 1986) appears to be better for glucose-6-phosphate dehydrogenase (E.C. 1.1.1.49) and adenylate kinase (E.C. 2.7.4.3).

Some isozymes are better than others for population genetic studies; avoid isozymes if you are not sure of the genetic basis of the variation observed. Recently, Fisk and Daly (1989) observed unreliable results with eggs. If eggs were reared at 25°C, eggs, 1-day-old appeared to express only the maternal genotype; those eggs close to hatching, about 3-days-old, expressed the larval genotype. Electrophoretic patterns of eggs of intermediate ages were ambiguous because extra bands appeared on the gels that were not present in either parent. Also, esterase loci are particularly difficult because of the presence of null alleles in many species, alleles that code for nonfunctional protein, and because some esterases are inducible by diet.

The gene and genotype frequencies produced in an electrophoretic study can be analyzed in a number of ways and are discussed in detail by Daly (1989). Modern techniques for analyzing such data are preferred, rather than the arc-sin square-root transformation. Log-linear modelling (also called G-tests), using statistical software packages, such a GLIM®[1] (Payne, 1985) or GENSTAT (Payne, 1987), provide a powerful means of analyzing genetic differentiation in and between populations (Daly and Gregg, 1985).

Useful measures of genetic differentiation within and between populations include deviations from Hardy–Weinberg frequencies using Smith's H statistic (Smith, 1970; Futuyma, 1979), Wright's F-statistics (Wright, 1978), and Nei's genetic distance, D (Nei, 1978). These can be calculated with the software program BIOSYS (Swofford and Selander, 1989). Inferences about gene flow between populations may be drawn using conditional allele frequency distributions (Slatkin, 1981).

However, these measures need to be used with caution (Daly, 1989). Contingency chi-square tests and G-tests are not necessarily appropriate for analysis of temporal variability, although they are used widely in this way (Waples, 1989). Estimates of F-statistics are sensitive to the number of populations and individuals sampled and the heterozygote frequencies (Weir and Cockerman, 1984). Genetic distance measurements can have large standard errors, particularly for small distance values (Richardson et al., 1986), whereas Slatkin's (1981) models have difficulty distinguishing between historical and contemporary patterns of gene flow (Caccone, 1985).

Taxonomic studies can use electrophoretic variation to distinguish different species, and sometimes, biotypes may be detectable. The most useful loci are those that are fixed for alternative alleles; even closely related species can differ in this way for up to 50% of loci. Interspecific differences are relatively easier to study than are intraspecific differences because the genetic differentiation is much greater. Consequently, the required sample sizes needed are smaller (Richardson et al., 1986). Genetic differences between such taxa as populations, subspecies and species, and genera, can be summarized using a number of statistics such as Nei's genetic distance, D (see above), Rogers (1972) genetic similarity, I, and various hierarchical cluster analyses (such as UPGMA and WPGMA), which are reviewed by Richardson et al. (1986) (Sneath and Sokal, 1973; Felsenstein, 1982). Again these analyses can be performed by BIOSYS. Electrophoretic data, however, may be inappropriate to infer phylogenetic relationships as the genetic distance between the taxa increases (Richardson et al., 1986).

2. Cytology

Cytological studies undertaken with *Heliothis* spp. are by Chen and Graves (1970); Laster (1972); Laster, Goodpasture, King, and Twine (1985); and

---

[1]GLIM® is the registered trademark of the Royal Statistical Society.

Fisk (1989). It is difficult to visualize genetic variation in the karyotype (the complement of chromosomes in each cell) because it is hard to differentiate between individual pairs of chromosomes. Members of the genus have 31 pairs of small holocentric chromosomes (chromosomes with a diffuse centromere rather than the normal single centromere). Before embarking on a cytological study of *Heliothis*, it would be necessary to demonstrate that variation did exist between individuals. Differences between chromosomes and between individuals can be visualized in some species by selectively staining parts of each chromosomes to produce C-bands, which are regions of heterochromatin. J. Fisk (personal communication), however, was unable to detect any C-bands in her study of *Heliothis* spp.

Fisk (1989) describes the methods for observing meiosis and mitosis in *H. armigera* and *H. punctigera*. Eggs midway through development were the best stage to observe metaphase in mitosis. At this time, the egg membrane develops a ring of yellow pigment. If eggs are kept at 25°C, then this stage is reached when the eggs are 1–1.5-days-old. In males, the first metaphase of meiosis was found in final larval instar. At 25°C, larvae are 13–15-days-old at this time. Because most of the cells in the testis go into meiosis in synchrony, many cells from each individual can be observed. Female meiosis was observed in sectioned material from ovaries of adults 2 days after emergence. C. Goodpasture (personal communication) reported some variation in the number of supernumerary chromosomes in *H. armigera* collected from Emerald in Queensland, Australia.

There are many different techniques for studying the karyotype of insects but no general guide appears to be available. This may be because techniques have to be modified for each new species. Examples of karyotypic studies of lepidopterans both at meiosis and mitosis are given by Fisk (1989). For a cytogenetics study of *Heliothis* you will need a good quality light microscope with a ×100 objective lens.

3. Molecular Biology

Only a few studies have been reported on the molecular biology of *Heliothis* spp. (Miller, Huettel, Davis, Weber, and Weber 1986; Davis and Miller, 1988) but none have dealt with the genetics of populations. If you wish to look at genetic variation at the level of the DNA, you would need to work cooperatively with a molecular biologist in their laboratory. A very useful laboratory manual is that by Maniatis, Fritsch, and Sambrook (1982). Applications of molecular techniques to evolutionary questions can be found in Avise, Lansman, and Shade (1979); Kreitman and Aguade (1986); Loukas, Delidakis, and Kafatos (1986); and Solignac and Monnerot (1986). Analysis of molecular variation provides a very powerful tool, but first consider if electrophoretic variation can be used to answer your question.

4. Morphological Variation

Differences in the external morphology of an insect can be used to detect differences between populations. It is necessary to establish that the character, such as variation in wing color, is determined genetically and not by environmental conditions. A standard set of crosses are performed in which the two color morphs are crossed and then the F_1 progeny are backcrossed to each parent. If the insects are reared under identical conditions, and the genetic basis of the character is relatively simple, then the proportions of each color morph should agree with Mendelian ratios for one or more genes. Once the genetic basis of the character has been established, it may be possible to analyze the differences among populations as illustrated above.

5. Insecticide Resistance

Insecticide resistance in field populations usually is determined by a single gene (Whitten and McKenzie, 1982; Roush and MacKenzie, 1987; Roush and Daly, 1990) although polygenic inheritance for resistance is always present, even in field-collected susceptible populations. The first step in a population genetics study of insecticide resistance is to determine the LD_{99}, the dose which kills 99% of susceptible (SS) individuals. This dose is obtained by screening a susceptible colony of *Heliothis* with a range of concentrations of the insecticide. Dr. R. Teakle, Queensland Department of Primary Industries, Indooroopilly, currently maintains susceptible colonies of *H. armigera* and *H. punctigera* (see Chapter 3).

To establish strains of *Heliothis* spp. that are homozygous (called RR) for the gene for resistance, we bring insects into the laboratory from the field and test them with the discriminating dose. If a single gene is involved, then the survivors should be a mixture of heterozygotes (RS) and homozygotes (RR). A full probit analysis (Finney, 1971) of progeny of the F_2 (a cross between $F_1 \times F_1$) should yield the LD_{99} for the heterozygous genotype, which if there is good discrimination between the three genotypes, is the does at which 75% of individuals are killed; the probit line should have a distinct plateau at this point. The F_3 colony ($F_2 \times F_2$) can be tested with the higher discriminating dose to yield a strain that is RR.

This technique is not simple because it can be difficult to rear *Heliothis* strains that have been selected with insecticides. Also, if the probit lines for the three genotypes overlap, then the true discriminating doses do not exist. The situation is complicated further if there is more than one gene for resistance. We have managed to establish homozygous strains for pyrethroid resistance with the use of isofemale lines. In this technique, single pairs of individuals from a resistant strain were mated and their progeny bioassayed at a full range of doses. The progeny from each isofemale line (brothers and sisters) were mated only with progeny from that line to yield an F_2 and F_3 generation. Isofemale lines whose dose-response curves were linear and parallel to that for the susceptible colony were considered homozygous-resistant. This can be confirmed by making single pair crosses of this new resistant strain with a susceptible strain. All crosses should yield a straight F_1 line parallel to both parents.

Once a homozygous strain has been established, this is crossed then to the susceptible colony and the progeny of each cross is bioassayed. The initial cross is SS × RR if resistance is monogenic. This involves two crosses to allow for the males and females of each genotype to be crossed with the reciprocal genotype (that is, SS × RR or RR × SS where the first genotype is that of the female parent). The progeny from these crosses should all be heterozygous, RS. These are backcrossed then to the susceptible parent (SS) again in both directions (RS × SS and SS × RS). The shape of the dose-response curve, particularly in the backcross, indicates the number of genes involved in resistance and also the degree of dominance (Tsukamoto, 1963, 1983; Roush and Daly, 1990). Care needs to be exercized in the interpretations of such probit lines, however, because misleading or ambiguous results can occur if there is not clear discrimination between genotypes (Roush and Daly, 1990).

If more than one gene is involved in resistance, then it is necessary to determine how much each gene contributes to the differences between individuals. Isofemale lines are established from the backcross generation. The progeny are screened using a variety of such techniques as electrophysiology or bioassays with synergists to determine which gene is represented in each isofemale line. Lines containing only one resistance gene can be retained for further study.

Once distinct genotype strains have been established and characterized, it is possible to measure gene frequencies in field populations directly rather than use indirect methods discussed above. This is essential if hypotheses about the evolution of resistance are to be examined. Various techniques to monitor resistance in field populations of *Heliothis* spp. are described by Forrester and Cahill (1987); Luttrell, Roush, Ali, Mink, Reid, and Snodgrass (1987); and Plapp et al. (1987) (Chapter 12). Daly, Fisk, and Forrester (1988) report on a method to measure selection for resistance in field populations.

Chapter 14

Modelling

R.L. Kitching

I. Introduction

The use of models in the study and management of pests has become commonplace in insect ecology. There remains among applied entomologists, however, an attitude that this is a specialist and separate task to be left to "someone else". Of course, a team approach to problem-solving may well permit use of specialists of this kind, just as it might include those with skills in toxicology, physiology, or applied economics. More often than not, insect ecologists and pest managers must work alone or in small nonspecialist teams. It is essential that such workers have an appreciation of the role of models, and an ability to build such constructs themselves — at least up to the stage of model development where technical execution takes over from conceptualization.

It is beyond the scope of a short account of this kind to provide all or even a large part of the technical skills required for the model-building process. I have written elsewhere at length on this topic (Kitching, 1983). Instead, I shall describe the nature of models, the steps involved in their construction and what they can and cannot do. I review, briefly, existing models of *Heliothis* and, in addition, I examine two levels of modelling that either have been, or could be, applied to *Heliothis* species.

This chapter, of course, should be read in conjunction with others that focus more on primary data-collection techniques. In particular, the accounts of life table construction (Chapter 6), estimation of dispersal (Chapter 10), and the measurement of basic developmental parameters and relationships (Chapter 8) will be particularly apposite to any modelling effort.

II. The Purpose, Practice and Limitations of Simulation Models

We all build models. We observe part of the world, simplify, and abstract it in our minds and use that construct to attempt to solve problems as diverse as "How does the national economy work?" and "How can I make the car go?" In ecology, this problem-solving process absorbed the availability of modern computing devices with some enthusiasm. For the first time, a tool was available for the generation of hypotheses that could match the problems in complexity. Like many such tools — radiotracers and bomb calorimeters spring to mind — model building was somewhat abused early in its ecological history and over-enthusiastic claims were made about its capabilities. This period has passed now and we can use the techniques along with many others when and where appropriate.

Model building serves two functions in ecology: first, it generates hypotheses about processes, populations, or communities to be tested by experimentation and rejected or accepted pro tem in the standard scientific fashion; or, second, it may synthesize what we know about a phenomenon in such a way that problems involving that phenomenon can be tackled and, sometimes, solved. The steps involved are listed in Table 14.1 (Kitching, 1983), most are self-evident, but I shall dwell briefly on some of them.

1. Problem Definition

This most crucial stage is the one that was overlooked most often in early model-building efforts. Essentially, it answers the question: just what am I trying to do and why is a model the way to go? Models may be built to answer very specific simple questions or suites of questions of increasing generality. In the *Heliothis* context, we may wish to model the dose-response curve of the pest on a particular crop for a particular insecticide, or we may

Table 14.1. Steps in the Building of a Simulation Model[a]

1. Problem definition
2. Systems identification
3. Decisions on model type
4. Mathematical formulation
5. Decisions on computing methods
6. Programming
7. Parameter estimation
8. Validation
9. Experimentation

[a]From Kitching, 1983.

wish to model the whole life-system of the species in a region to evaluate integrated management strategies, one against the other. The key point is that all subsequent decisions and stages in the model-building process reflect this first step — it must be carried through with clarity of thought and a strong dose of pragmatism: no model is a panacea for all problems associated with a pest system nor, incidentally, can it make up for inadequate biological knowledge (or any other sort of knowledge, for that matter!).

2. Systems Identification

Building a simulation model is an exercise in systems analysis. As in any such analysis, we define the system to match the question being asked. We identify measurable state variables (e.g., the numbers of larvae, eggs, pupae, or what have you, in a population) that we need. We define sources and sinks of material (e.g., pools of migrating adults elsewhere, or detritus pools to which dead organisms are added). We connect these by material and information flows (e.g., transfers of material as eggs become larvae, or the influence of adult density on survivorship). These flows imply the existence of rate processes such as hatching, development, death, emergence, and so forth that produce the dynamic changes we observe in our state variables. Together with a few other simple notions such as driving variables and constants these define the system with which we shall work. Figure 14.1 is a schematized version of a simple system. Note that the *system* does not exist until we define it (in contrast with the assumptions behind the plain English use of the term). Ideally, what we include in the system should be no more nor less than we need to tackle our selected problem set. Excessive detail is as disastrous as is too little in this regard.

3. Parameter Estimation

Once we have defined our system, we can write simple equations describing the changes that occur through time in each of our state variables. This set of state equations is the basis of our model. However, to go from such simple equations of change (which essentially add and subtract material from "past" values of each state variable) we need to know how fast each rate process operates within a particular time period. This demands knowledge of how the flows change with respect to the current values of the various entities that have information connections with the rate variables involved. The mathematical formulation of these relationships involves the estimation of parameter values within whichever statistical relationship we choose (e.g., the a and b in the regression model: $y = a + bx$). Once we reach the point of estimating parameter values, our model becomes one of a particular species, population, or crop. Such parameter values are obtained in one of three ways. For a pest model these might be, in descending order of desirability:

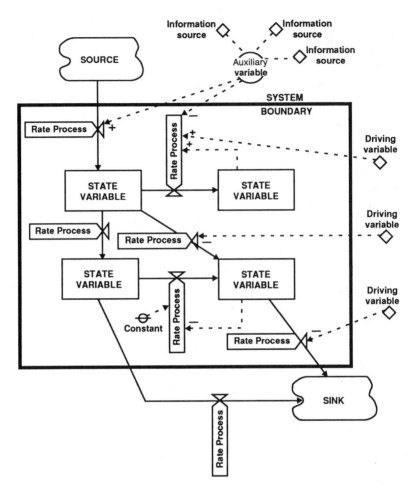

Figure 14.1. A hypothetical system showing some of the characteristic components. State Variables, quantities (such as population numbers) that can be measured at any point in time; Rate Processes, processes that cause changes in the levels of state variables through time (for example, mortality, reproduction, immigration, and emigration); Driving variables, factors that affect the rate of operation of such processes, but that operate from outside the system as defined; Constants, fixed values that may affect also the rate of operation of driving variables; Sources and Sinks, pools of material that feed into or receive material from state variables; material flows (solid lines), the actual movement of material within the system and across the System Boundary; information flows (dashed lines), influences on the rate of operation of rate processes that sometimes are grouped conveniently into named Auxiliary variables.

1. Values based on statistical analysis of data collected on the subject species at the location of interest over one or more seasons.
2. Values obtained from the literature on the subject species from elsewhere or, even on related species.
3. Values that are informed guesses based on the modeller's experience or intuition.

Obviously, too great a proportion of Type (3) values will introduce serious doubts about the credibility of the model, although too moralistic an approach may result in no model ever being built. Once a first-pass set of values is obtained the model can be subjected to a sensitivity analysis. In essence, this tries to determine how sensitive are the predictions of the model overall, to changes in the values of particular variables (see, e.g., Steinhorst, Hunt, Innis, and Haydock, 1978). If such an analysis suggests great sensitivity to one of the Type (3) parameters, then substantial additional effort to obtain a more reliable value is warranted. If the model is insensitive to these parameters, then perhaps their basically inadequate nature can be lived with, at least for the time being.

Parameter estimation, implying as it does, the need for an extensive data base is the most time-consuming step of the whole model building exercise. Ideally, data collection and analysis should be done in parallel with the model-building process — sadly, this is seldom possible.

4. Validation

Once a simulation reproduces accurately the values and relationships that were built into it explicitly, that is, the model has been verified, then its predictive powers must be tested. This process is called *validation* and it is carried out with more or less vigor depending on the problems being tackled and the opportunities and biases of the workers involved. For some purposes, an "adequate" correspondence between prediction and suitable test data (that is, data not used in the parameter estimation stage of the work) may be an approximate reproduction of the major peaks and valleys in the test data: for other purposes, a difference of more than 10% say, may be unacceptable. In either case an unvalidated model is worthless, and the breadth of appropriate application, even of a validated one can be established only in the light of the validation procedure that was applied. Each user of a model must satisfy themselves that it is validated for the particular intended use, which may be different from that of the original modeller.

5. Experimentation

This is the term used for the actual application of the model. It cannot be stressed too strongly that a model is at most an interim research goal in an entomological program. In any case, it is the experimentation phase of the work that turns the whole enterprise into science rather than into one solely of artistic creation.

Simulation modelling can help the ecologist achieve a number of things: it will synthesize information about a system thereby identifying significant gaps; it will make predictions about the likely effects of ecological manipulations; and, it can provide stand-alone general constructs that add to the ecological theory about processes, populations, communities, or ecosystems. However, there are certain things that the modelling process most certainly cannot do. No model can substitute for good data on the subject system: indeed, it is dependent wholly on such data for its construction, validation, and use. It is risky to use a model for some purpose not stated by the builder, and such extended use demands extensive rechecking and validation. Models are not all-powerful and it should not be assumed that some unspecified modelling effort will solve problems unamenable to other techniques.

All this having been said, models have proved enormously useful in insect ecology and management. These models fall into four classes:

1. Simple algebraic models of ideal populations (see e.g., May 1973, 1981, 1986; Roughgarden, 1979).
2. Complex simulations of ecological processes such as predation, competition, and movement (Holling, 1964,1966; Griffiths and Holling, 1969; Kitching and Zalucki, 1982; Zalucki and Kitching, 1982b).
3. Life-system models of single species, their hosts and natural enemies (Gutierrez, Havenstein, Nix, and Moore, 1974; Geier and Clark, 1976; Gilbert, Gutierrez, Frazer, and Jones, 1976).
4. Decision-making and tactical models at the crop or regional level (Room 1979; Hearn and da Roza, 1985).

III. Existing Models for *Heliothis* Ecology

For such an important pest species, relatively few models have been constructed for aspects of the ecology, behavior and impact of *Heliothis* to date. Most of these have been North American efforts although the SIRATAC cotton-management model is an Australian development of some consequence.

A number of authors have modelled selected segments of the biology of *Heliothis* species. Butler and his co-workers (Butler, 1976; Butler and Hamilton, 1976; Butler, Hamilton, and Proshold, 1979) modelled the temperature-development relationships for *H. subflexa*. Eger (1981) is reported by Hartstack (1982) to have modelled temperature-induced overwintering mortality of *H. virescens* and *H. zea*, an area examined more analytically by Logan et al. (1979) for *H. zea*. Stinner and his co-workers (1974b) simulated the population dynamics of *H. zea* and *H. virescens* using temperature-

dependent developmental rates, responses of adults to host-plant characteristics, and cannibalism among larvae. Using this model, they were able to
predict peaks of activity that coincided well with light-trap data although
the sizes of the peaks were less well imitated.

Beginning in 1973, A.W. Hartstack and his colleagues produced a series
of "complete" population models for *H. virescens* and *H. zea*, which they
named *MOTHZV*, *MOTHZV-2*, and *MOTHZV-3*. The first of these (Hartstack et al., 1973; Hartstack and Hollingsworth, 1974) used the numbers of
adults emerging from diapause, subsequently modified by ambient temperatures, moon phase, and crop phenology to predict the size and timing of
light-trap peaks. It was a location-specific, regression-based formulation,
highly successful for its place of construction (College Station, Texas) but
less so elsewhere. MOTHZV-2 was a considerably more sophisticated version, represented in Figure 14.2 (modified from Hartstack, Witz, Hollingsworth, Ridgeway, and Lopez, 1976). The age-structured dynamics of
the moth represented by the model included the driving effects of temperature, natural enemies, insecticide use, cannibalism, and three alternative sets
of host-plants — corn, cotton, and sorghum. MOTHZV-3 added a model of
Heliothis damage to fruiting cotton to replace the simple crop effects incorporated into MOTHZV-2 (Hartstack, 1982).

Turning to tactical models for particular crops, I note that cotton has been
a focal point to date. Brown, McClendon, and Jones (1979) and Jones,
Brown, and Hesketh (1980) constructed models of cotton crops and their
major pests based in Mississippi, and Wang, Gutierrez, Oster, and Daxl
(1977) went some way along this route for acala cotton in California. Hartstack (1982) reports on BUGNET, a multicrop tactical model for *H. zea*
based in Texas and, of course, the multi-institutional team based in the
Namoi, Australia, have developed SIRATAC.

SIRATAC has developed through a series of versions during the last 10
years, building on the prototype published by Room (1979), each adding
submodels, greater flexibility, and capability to the original. Figure 14.3,
provided by Mr. Brian Hearn of the CSIRO, summarizes the model components . The program uses agronomic and climatic data, together with results
of insect and fruit counts to assess likely damage and economic loss. It
contains a *Heliothis* development model, a model of cotton fruiting, and a
model of *Heliothis* feeding. Threshold values for damage are used in conjunction with the biological modules to make management recommendations through a so-called expert system incorporated in the whole (Stone,
Coulson, Frisbie, and Loh, 1986). The use of SIRATAC has spread rapidly,
and in the 1984–1985 season was used to manage 27% of the national cotton
crop. General discussions of the model and its implementation at various
stages are provided by Brook and Hearn (1983), Hearn and Brook (1983),
Ives et al., (1984), and Hearn and da Roza (1985).

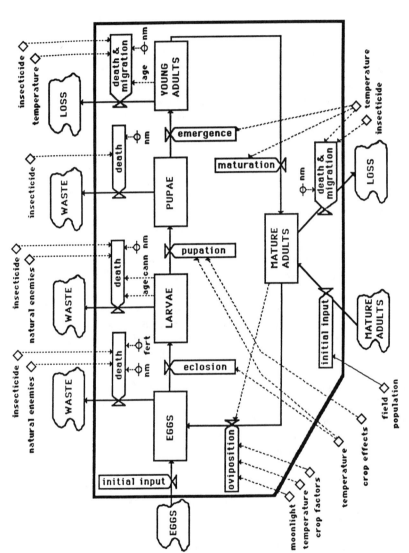

Figure 14.2. Forrester diagram of the population model MOTHZV-2 of Hartstack et al. (1976). Abbreviations: nm, natural mortality; cann, cannibalism; fert, fertility. Variables listed around the edge of the diagram are those used together with some internal constants and variables by Hartstack and his colleagues to calculate the magnitude of the rate variables (drawn by R.L. Kitching).

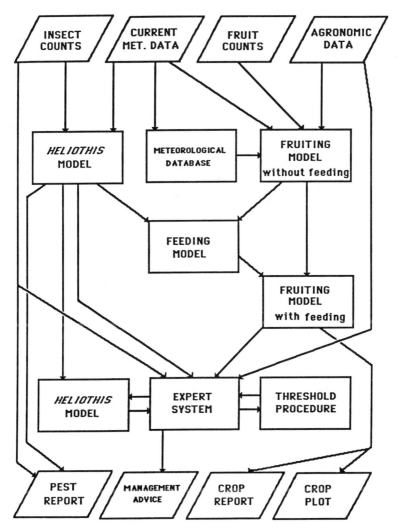

Figure 14.3. The major components of the SIRATAC management model. The sloped boxes represent inputs and outputs, the rectangular structures submodels or databases. (Drawn by R.L. Kitching with permission from Mr. Brian Hearn.)

IV. Selected Modelling Methods for *Heliothis* Populations

I have chosen to present below an outline of two modelling methods that allow representation of *Heliothis* dynamics at either the simple population or life-system levels. Each has great potential for application in a variety of problem-solving situations and place different levels of demand both on modelling skill and the completeness and detail of the requisite data base.

1. Leslie Matrices

The simplest sort of population model we can build, examines the additive and subtractive processes that are operating on the population (suitably defined in time and space) in any time period. In its simplest form, we have:

$$\Delta N = B - D + I - E$$

where ΔN is the change in the population, B is births, D deaths, I immigrants, and E emigrants, over the time period of interest. This equation is the basis of many simple population models and, indeed, with only a little added detail can produce models of surprisingly complex and interesting dynamics (May, 1981,1986). In studying insects, we are interested almost always in populations that have age structure: at the very least, we want to be able to represent eggs, larvae, pupae, and adults separately. A more complex, but very useful mathematics is required for this purpose: linear algebra. In its simplest form such a representation would be as that in Table 14.2. The f_i terms represent the fecundity of individuals in each age-class, that is, the number of offspring born to individuals of each age within a fixed time period. Strictly these terms and others within the matrix and vectors involved, refer to the females within a population although they conveniently can be thought of as across-sex averages if required. Indeed, it must be stressed that the model represents nothing more than a dynamic version of the standard age-specific life table with an added fecundity schedule. Once having constructed the equation and decided on the size and structure of our initial population, we can predict the size and structure of the population at intervals into the future by carrying out simple matrix multiplications as shown in Table 14.3. The matrix containing the *fecundity* and *survivorship* terms is called a *Leslie matrix* after one of its originators (Leslie, 1945). This sort of model has a number of properties that make it particularly valuable. Most useful among these is the ability to identify the stable age structure of the population that is arrived at after a number of iterations of the Leslie model. Once this is arrived at, the discrete rate of

Table 14.2. Basic Form of a Leslie Matrix[a]

$$
\begin{bmatrix}
f_0 & f_1 & f_2 & \cdots & f_{k-1} & f_k \\
p_{0,1} & 0 & 0 & \cdots & 0 & 0 \\
0 & p_{1,2} & 0 & \cdots & 0 & 0 \\
\multicolumn{6}{c}{\dotfill} \\
0 & 0 & 0 & \cdots & p_{k-1,k} & p_k
\end{bmatrix}
\bullet
\begin{bmatrix}
n_0 \\
n_1 \\
n_2 \\
\\
n_k
\end{bmatrix}_t
= \mathbf{n}_{t+\Delta t}
$$

[a] $\mathbf{n}$ is a vector representing the number of individuals in each of k age-classes, the f_i's represent age-specific birth rates, and the p_{ij}'s are the age-specific survivorship terms.

Table 14.3. Prediction of Size and Structure of a Population at Intervals into the Future[a]

The Leslie Matrix (L)					Population Vector at Time, t ($\mathbf{n}_t$)		Intermediate Step		Population Vector at Time, $t+1$ ($\mathbf{n}_{t+1}$)
$\begin{bmatrix} 0 & 0 & 20 & 5 \\ 0.2 & 0 & 0 & 0 \\ 0 & 0.4 & 0 & 0 \\ 0 & 0 & 0.4 & 0 \end{bmatrix}$				$\bullet$	$\begin{bmatrix} 100 \\ 25 \\ 5 \\ 2 \end{bmatrix}$	$=$	$\begin{bmatrix} (0 \times 100) + (0 \times 25) + (20 \times 5) + (5 \times 2) \\ (0.2 \times 100) + (0 \times 25) + (0 \times 5) + (0 \times 2) \\ (0 \times 100) + (0.4 \times 25) + (0 \times 5) + (0 \times 2) \\ (0 \times 100) + (0 \times 25) + (0.4 \times 5) + (0 \times 2) \end{bmatrix}$	$=$	$\begin{bmatrix} 110 \\ 20 \\ 10 \\ 2 \end{bmatrix}$

[a] If we divide a population into four age-classes of equal duration (of 1 time-step) and find that they have reproductive rates (that is, offspring per individual per time-step) of 0 (f_0 and f_1), 20 (f_2) and 5 (f_3) and probabilities of survival from one age-class to the next of 0.2 ($p_{0,1}$), 0.4 ($p_{1,2}$), and 0.4 ($p_{2,3}$), then if we start with an initial population of 100, 25, 5, and 2 in each class, we can model the dynamics of the population using simple matrix multiplication. $\mathbf{Ln}_t = \mathbf{n}_{t+1}$, where L is the Leslie matrix.

change (*g*) of the population can be calculated, and, from this, the intrinsic rate of natural increase (*r*), from the identity: $r = lng$.

If we wish to apply this sort of model to insect populations, we encounter a number of problems.

1. The time-step involved is constant and, in the simple form, all members of a particular age-class either must graduate to the next class or die in any interval (both processes are summarized in the p_{ij} terms in the matrix). Obviously this is unrealistic when we are dealing with stadia of unequal length.
2. *Survival* and *fecundity* terms are constant. In reality, they would respond to environmental temperature, levels of natural enemies, density-dependent effects, food quality, and so forth.
3. Emigration and immigration are excluded.

Each of these drawbacks can be circumvented with a little ingenuity.

We can incorporate nonprogression into the matrix by utilizing some of the zero elements as shown in Table 14.4. *Survival* and *fecundity* terms can be made dependent variables calculated afresh at each iteration (= matrix multiplication) or some other selected periodicity, on the basis of one or more independent variables that may have been demonstrated to be important in other work. Emigration and immigration can be incorporated by additional operations as laid out in Table 14.5 (note that the order in which the subcalculations of this augmented Leslie model are carried out is important and that the structuring of a program to run such a model, must be done with great care). As we incorporate each additional complication into the model, we increase the data demands of the model and reduce the mathematical advantages inherent in the simpler versions of the model. Indeed, in due course, this adding in of complications will produce a model that converges on the life-systems models described below. Table 14.6 summarizes the information required to implement models based on the Leslie matrix approach.

Once constructed a Leslie matrix model is used to obtain quick predictions of likely population trends and can be useful in making informed comparisons between likely population events across different crops, regions or species (each of which would have different f_i and p_{ij} terms).

2. Life-System Models

Life-system models (also called dynamic life-table models by Gilbert et al., 1976) were developed, principally in Australia, as a means of summarizing the population dynamics of insect pests and those segments of the environment with which they interact significantly. Early work focused on aphids (Hughes and Gilbert, 1968; Gutierrez et al., 1974) and the best documented work of such modelling activity is that of Gilbert et al. (1976) of the thimbleberry aphid, *Masonaphis maxima*, in western Canada. Other pests

Table 14.4. Basic Population Transition with Nonprogression[a]

$$
\begin{bmatrix}
0.2 & 0 & 20 & 5 \\
0.2 & 0.2 & 0 & 0 \\
0 & 0.4 & 0.4 & 0 \\
0 & 0 & 0.4 & 0.5
\end{bmatrix}
\cdot
\begin{bmatrix}
100 \\ 25 \\ 5 \\ 2
\end{bmatrix}
=
\begin{bmatrix}
(0.2 \times 100) + (0 \times 25) + (20 \times 5) + (5 \times 2) \\
(0.2 \times 100) + (0.2 \times 25) + (0 \times 5) + (0 \times 2) \\
(0 \times 100) + (0.4 \times 25) + (0.4 \times 5) + (0 \times 2) \\
(0 \times 100) + (0 \times 25) + (0.4 \times 5) + (0 \times 0.5)
\end{bmatrix}
=
\begin{bmatrix}
130 \\ 25 \\ 12 \\ 3
\end{bmatrix}
$$

[a]If the population described in Table 14.3 now has age-specific nonprogression rates of 0.2, 0.2, 0.4, and 0.5, respectively (that is, at each population transition 20% of age-class 1 remains in class 1, 20% in class 2, and so on), then we can model the basic population transition as described above. This structure copes with the problem of age-classes of different durations (such as insect stadia). Thus we have constructed an augmented Leslie matrix with the nonprogression rates added as the principal diagonal.

Table 14.5. Incorporation of Emigration and Immigration[a]

Reproduction and Mortality (L)	Population at Time, t $(\mathbf{n}_t)$	Immigration Vector $(\mathbf{n}_i)$	Emigration Vector $(\mathbf{e}_t)$	$\mathbf{n}_t$	Population at Time, $t+1$ $(\mathbf{n}_{t+1})$
$\begin{bmatrix} 0 & 0 & 20 & 5 \\ 0.2 & 0 & 0 & 0 \\ 0 & 0.4 & 0 & 0 \\ 0 & 0 & 0.4 & 0 \end{bmatrix} \bullet$	$\begin{bmatrix} 100 \\ 25 \\ 5 \\ 2 \end{bmatrix} +$	$\begin{bmatrix} 2 \\ 20 \\ 4 \\ 0 \end{bmatrix} -$	$[0.2\ 0.6\ 0.2\ 0] \bullet$	$\begin{bmatrix} 100 \\ 25 \\ 5 \\ 2 \end{bmatrix} =$	$\begin{bmatrix} 92 \\ 25 \\ 13 \\ 2 \end{bmatrix}$

[a]Consider now the population described in Table 14.3 to which must be added a number of immigrants at each transition and from which a number of emigrants departs each time. The number of immigrants is determined elsewhere (although they could be modelled in a multipopulation system), and the number of emigrants is a fixed proportion of each age-class, say 0.2, 0.6, 0.2, and 0, respectively. $\mathbf{n}_{t+1} = (L \cdot \mathbf{n}_t) + \mathbf{n}_i - (\mathbf{e} \cdot \mathbf{n}_t)$ where L is the Leslie matrix, $\mathbf{n}_i$ is the immigration vector, and $\mathbf{e}$ is the (row) vector of emigration rates.

have been approached in similar fashion, including bushflies (Hughes and Sands, 1979) and the codling moth (Geier and Hillman, 1971).

Basically, a model of this kind follows the age and/or morph structure of the population through a season, modifying numbers according to additive and subtractive processes that operate within any time period. Relatively unimportant processes may be incorporated using simple empirical relation-

Table 14.6. Information Required for the Construction of the Leslie Matrix Models

Nature of Model	Data Required
"Basic" recursion model	Estimates of survival rates over equal periods of the insect's life
	Measures of reproductive rates for each period of the insect's life
"Augmented" models	The above **plus**:
With nonprogression	Information on the durations of selected age-classes in terms of the time-steps selected earlier
With nonconstant elements	The relationships among survival and reproductive rates and selected extrinsic factors: climatic conditions, abundance of natural enemies, food quality, species density, control pressures, and so on.
With immigration and emigration	Regular estimates of the numbers and ages of immigrants
	Age-specific emigration rates[a]

[a]Note that such rates are unlikely to be constant and may have to be recalculated regularly on the basis of extrinsic factors such as population density, season, individual quality, and so forth.

ships, whereas other more important factors (either quantitatively, or simply in the perception of the modeller) may be included as more complex modules. Life-system models are complex and usually written in higher level or simulation programming languages (such as FORTRAN, PASCAL, SIMULA, DYNAMO, etc.), which reflect the time-based logic defined by the modeller. Essentially, models of this kind operate on a central time-loop during which the various population mechanisms are simulated, generating a "new" population structure. This iterative process is continued until a complete season (or more) has been simulated. The similarity in principle to Leslie matrix models is not accidental, but the techniques of modelling involved are far more wide-ranging and less constrained by the demands of the mathematics involved — in one sense, life-system models are biologists' models rather than borrowings from mathematicians (who are often horrified by their analytical intractability and general "messiness").

Life-system models of insect pests demand a common core of information and techniques.

1. The temperature-development relationships of the subject species must have been worked out so that real time can be converted by thermal summation to degree-days (or vice-versa). This is an exercise of greater or lesser complexity depending on the range of temperatures used, the importance of complicating covariates (such as diet, size, or sex), the effects of fluctuating temperatures, and the role of solar radiation (see Chapter 8). Some authors have converted degree-days into more useful (for them) units such as the Quarter Instar Periods (QUIPS) of Gilbert and Gutierrez (1973) and Gutierrez et al. (1974). In these models, this carried the implication that those parts of the simulation model that update age-structures are entered only when at least one QUIP's worth of thermal input has been accrued. There are many accounts of the techniques and principles behind such age-grading techniques. I have discussed them in Kitching (1977), and Tyndale-Biscoe (1984) has reviewed the morphological aspects of age-grading techniques.

2. Information on age-specific reproductive rates and the ways in which these respond to differing environmental circumstances is essential. Here the role of adult-size in determining fecundity can be incorporated as can the impacts of food quality, post-teneral flight requirements, reproductive diapause, factors influencing oviposition, and so forth. As with many aspects of life-system modelling, once the need for key relationships has been established, the actual relationships incorporated into the model may be determined by regression analysis, or by incorporating actual observations that are, thereafter, used as a pattern for prediction. FORTRAN data statements, for example, lend themselves well to this last purpose, as do the various algorithms that allow simulation of arbitrarily defined distributions (Butler, 1970).

3. Information on mortality processes is also vital, complimentary as it is to the additive reproductive processes described above. The amount of detail that can be incorporated at this point is well-nigh limitless and, as

always in a modelling exercise, must represent a compromise between the demands of a particular problem and the available data and/or research resources. In general, a basic ("intrinsic") survivorship curve for the species is obtained in the laboratory in which the longevity and mortality patterns of well-fed individuals are measured, often over a series of temperatures. Superimposed on this are estimates of the additional mortality due to overcrowding (including cannibalism), food quality, edaphic, and climatic factors, and, of course, the impact of natural enemies — predators, parasitoids, and pathogens. Again, the choice rests with the particular worker as to whether she or he uses simple empirical estimates, first-order statistical relationships, or more or less complex structural algorithms to represent each of these processes. In practice, few have found it appropriate or feasible to do more than incorporate simple models of these phenomena. Gilbert et al. (1976), for example, included detailed information only on the principle predator of *Masonaphis maxima*. Inclusion of even this level of detail required information on developmental rates and temperature relations of the syrphid concerned, its average voracity across all larval instars, and its basic field phenology.

4. In many species (and species of *Heliothis* fall squarely into this category), immigration and emigration are crucial processes in the understanding of the dynamics of the populations of interest. Immigration is probably best regarded as a driving variable (Fig. 14.1) and empirical data from trap catches supplied at each iteration. Nevertheless, information will be required about the age-structure of the immigrants and their reproductive status. Emigration is a more difficult problem. Individuals may be lost to the population on a regular basis across age-classes, periodically for a particular age-class, or in response to subtle density-dependent pressures on an age-class or some precursor stage of that age-class.

5. Lastly, management and/or damage factors can be incorporated into life-system models. In many instances, management options can be added to existing mortality or reproductive structures within a model: for instance, artificially increasing the densities and, consequently, the impact of natural enemies (as in biological control), by reducing the numbers of particular age-classes (as in chemical control), by reducing the reproductive responses to food intake (as in cultural control) or by decreasing average fecundity (as in genetical control).

The above five key modules may be added to in response to particular situations. They may be embedded in a spatial model of a crop mosaic, for instance, or collapsed into a "pest" module of a model of a particular crop.

My account of life-system models hereto has been general. Turning to such models of *Heliothis* spp., only the models of Hartstack (1976) for *H. virescens* and *H. zea* are of this kind and they are far from complete in terms of the above criteria. Satisfactory models at this level for the Australian situation will require a stronger data base than is currently available. I have summarized in Table 14.7, the classes of data that are required for this purpose, although of course, special-purpose models (relating say, to particular crops, seasons, regions, or pest species) may be built with less information.

Table 14.7. Information Required for Construction of a Life-System Model for an Insect Pest

Module	Information Required
Development rates	Stage durations at a range of constant temperatures
	Effects of fluctuating temperature and/or solar radiation
	Effects of foodplants on development
	Ambient temperatures for periods and sites of interest
Reproduction	Effects of larval and adult nutrition on fecundity
	Relationships between fecundity and adult size
	Age-specific changes in reproductive rates
	Density-dependent effects on fecundity
	Effects of environmental factors (e.g., temperature, rainfall, etc.) on fecundity
	Effects of availability of mates
Mortality	Intrinsic physiological mortality curves
	Effects of individual quality on survivorship
	Temperature effects
	Density-dependent effects due to crowding, cannibalism, and so on
	Density-independent effects due to climate or other physicochemical changes
	Natural enemy dynamics
Movement	Phenology and age-structure of immigrants in sites and seasons of interest
	Effects on emigration
	Density-dependent inducement
	Density-independent conditions
Control	Point of impact and magnitude of effect (or expected effect) of each of the variety of control measures
	Costs and benefits of control measures and impacts
	Magnitude and nature of interactions among control measures

V. Concluding Remarks

Simulation models are responses to needs associated with the understanding of complex objects. *Heliothis* spp. and the syndrome they engender comprise such a complex object and we should expect models of this kind to play an important role in understanding and managing the problems involved. The basic structures of models at a variety of levels have been described here. They are not new and have been executed many times by many authors with respect to a variety of other organisms. Those concerned with the study and management of *Heliothis* will be able to add to these structures on the basis of their understanding, experience, and intuition. Equally, they will recognize the data base required to implement these models further. Herein lies the rub: each part of each model requires particular information that frequently will exceed that which is currently available even for such "simple" models as the Leslie matrix structure. Accordingly, the most useful route that such efforts in "conceptual" modelling (with their simple formalization as systems diagrams) may lead us along, is as definers of programs of data collection. They tell us what we do not know within that set of things we need to know, which in turn reflects the aims of our work in a broader context.

This having been said, it is more difficult to build useful models once the data-collection process is complete: refining and improving upon earlier structures, yes, but starting the modelling exercise at that point will always be unsatisfactory. I suggest that instead of regarding the modelling enterprise as an option in any program of ecological study, it should be incorporated from the outset as the framework within which all other data are gathered. There is nothing worse than trying to fit together the pieces of a jigsaw only to find at the end, that some crucial pieces are missing!

Acknowledgments. This work would have been impossible without the very useful discussions I had with participants in the *Heliothis* workshop held in Brisbane, Australia in July 1985 and I am grateful to all those concerned. In addition, key pieces of literature, both published and unpublished, were brought to my attention by Drs. Myron Zalucki and Peter Allsopp. Lastly, Brian Hearn of the CSIRO, Narrabri, Australia, has been particularly generous in making available to me unpublished material on SIRATAC in all its stages. He also allowed me to reproduce Figure 14.3 from among unpublished material.

References

Adams JB, van Emden HF (1972) The biological properties of aphids and their host plant relationships. In: HF van Emden (Ed), *Aphid Technology*. New York: Academic Press, pp.48–104.

Agee HR (1973) Spectral sensitivity of the compound eyes of field-collected adult bollworms and tobacco budworms. Ann Ent Soc Am 66:613–615.

Allendorf FW, Phelps SR (1981) Use of allelic frequencies to describe population structure. Can J Fish Aquat Sci 38:1507–1514.

Allsopp PG (1986) Development, longevity and fecundity of the false wireworms *Pterohelaeus darlingensis* and *P. alternatus* (Coleoptera: Tenebrionidae). IV. Models for larval and pupal development under fluctuating temperatures. Aust J Zool 34:815–825.

Allsopp PG (1988) Modelling development of larvae and pupae of *Pterohelaeus darlingensis* Carter with Podolsky phenocurves. Aust J Ecol 13:411–413.

Andrewartha HG (1952) Diapause in relation to the ecology of insects. Biol Rev 27:50–107.

Annecke DP, Moran VC (1977) Critical reviews of biological pest control in South Africa.1. The Karoo caterpillar, *Loxostege frustalis* Zeller (Lepidoptera: Pyralidae). J Ent Soc Sth Afr 40:127–145.

Anon (1970) Standard method for detection of insecticide resistance in *Heliothis zea* (Boddie) and *H. virescens* (F.). Bull Ent Soc Am 16:147–153.

Anon (1983) Pyrethroid resistance. The Aust Cotton Grower 4 (3):4–7.

Archie J, Simon C, Wartenberg D (1985) Geographical patterns and population structure in periodical cicadas based on spatial patterns of allozyme frequencies. Evolution 39:1261–1274.

Arn HE, Stadler E, Rauscher S (1975) The electroantennographic detector – a selective and sensitive tool in the gas chromatographic analysis of insect pheromones. Z Naturforsch C 30:722–725.

Ashby JW (1974) A study of arthropod predation of *Pieris rapae* using serological and exclusion techniques. J Appl Ecol 11:419–425.

Avise JC (1974) Systematic value of electrophoretic data. Syst Zool 23:465–481.

Avise JC, Lansman RA, Shade RO (1979) The use of restriction endonucleases to measure mitochondrial DNA sequence relatedness in natural populations. I. Population structure and evolution in the genus *Peromyscus*. Genetics 92:279–293.

Ayala FJ, Tracy ML, Hedgecock D, Richmond RC (1974) Genetic differentiation during the speciation process in *Drosophila*. Evolution 28:576–592.

Bacon JP, Altman JS (1977). A silver intensification method for cobalt-filled neurons in wholemount prepartions. Brain Res 138:359–363.

Bacon OG, Seiber JN, Kennedy GG (1976) Evaluation of survey trapping for potato tuberworm moths with chemical baited traps. J Econ Entomol 69:569–572.

Baker TC (1986) Pheromone-modulated movements in flying moths. In: TL Payne, MC Birch, CEJ Kennedy (Eds), *Mechanisms in Insect Olfaction*, Oxford: Clarendon Press, pp. 39–48.

Baker TC, Carde RT (1984) Techniques for behavioural bioassays. In: HE Hummel, TA Miller (Eds), *Techniques in Pheromone Research*. Springer Series in Experimental Entomology, New York. Springer-Verlag, pp. 45–74.

Bartlett AC, Raulston JR (1982) The identification and use of genetic markers for population dynamics and control studies in *Heliothis*. In:W Reed,V Kumble (Eds) *Proceedings of the International Workshop on Heliothis Management*, November 15–20, ICRISAT Center, Patancheru, A.P., India.

Bateman MA (1976) Fruit flies. In: VL Delucci (Ed), *Studies in Biological Control*. Cambridge: Cambridge University Press, pp.11–50.

Bechinski EJ, Pedigo, LP (1982) Evaluation of methods for sampling predatory arthropods in soybeans. Env Ent 11:756–761.

Beck SD (1983) Insect thermoperiodism. Ann Rev Entomol 28:91–108.

Bedford ECG (1951) Unpublished report. Pretoria: Plant Protection Research Institute.

Bedford ECG (1956) The automatic collection of mass-reared parasites into consignment boxes, using two light sources. J Ent Soc Sth Afr 19:342–353.

Begon M (1979) *Investigating Animal Abundance: Capture-Recapture for Biologists*. London: Edward Arnold.

Begon M, Mortimer M (1981) *Population Ecology*. Oxford: Blackwell Scientific.

Bell JV (1969) *Serratia marcescens* found in eggs of *Heliothis zea*. Tests against *Trichoplusia ni*. J Invertebr Pathol 13:151–152.

Bell JV (1975) Production and pathogenicity of the fungus *Spicaria rileyi* from solid and liquid media. J Invertebr Pathol 26:129–130.

Bell MR (1988) *Heliothis virescens* and *H. zea* (Lepidoptera: Noctuidae) feasibility of using oil soluble dye to mark populations developing on early-season hot plants. J Entomol Sci 23:223–228.

Bell WJ (1984) Chemo-orientation in walking insects. In: WS Bell, RT Carde (Eds), *Chemical Ecology of Insects*, London: Chapman and Hall, pp. 93–109.

Belton P, Kempster RH (1963) Some factors affecting the catches of lepidoptera in light traps. Can Ent 92:832–837.

Bent GA (1984) Developments in detection of airborne aphids with radar. In: *Proc 1984 Brit Crop Prot Conf Pests and Diseases*. Vol 2. Croyden: British Crop Protection Council, pp.665–674.

Berlocher SH (1979) Biochemical approaches to strain, race, and species identification. In: MA Hoy, JJ McKelvey Jr, (Eds), *Genetics in Relation to Insect Management*. New York: The Rockerfeller Foundation, pp.137–147.

Betts E (1976) Forecasting infestation of tropical migrant pests: the desert locust and the African armyworm. Proc Symp R Ent Soc Lond 7:113–134.

Bilapate GG (1981) Investigation of *Heliothis armiger* Hübner in Marathwada. XXIII. Key mortality factors on cotton, pigeonpea and chickpea. Proc Indian Acad Sci 47B:637–666.

Bilapate GG, Pawar VM (1978) Life tables for the gram podborer *Heliothis armigera* Hübner (Lepidoptera: Noctuidae) on pea. Proc Indian Acad Sci 87B:119–121.

Bilapate GG, Raodeo AK, Pawar VM (1981) Investigations on *Heliothis armigera* Hübner in Marathwada. III. Life tables and intrinsic rate of increase on chickpea. J. Maharashtra Agric Univ 6:51–54.

Bilapate GG, Raodeo AK, Pawar VM (1984) Investigations of *Heliothis armigera* Hübner in Marathwada. III. Life table studies on cotton squares. J Maharashtra Agric Univ 9:261–262.

Bishop AL, Blood PR (1980) Arthropod ground strata composition of the cotton ecosystem in south-eastern Queensland and the effect of some control strategies. Aust J Zool 28:693–698.

Bishop AL, Blood PR (1981) Interactions between natural populations of spiders and pests in cotton and their importance to cotton production in south-eastern Queensland. Gen Appl Ent 13:98–104.

Blanchard RA (1942) Hibernation of the corn earworm in the central and north-eastern parts of the United States. USDA Tech Bull No 838:1–13.

Blaney WM, Simmonds MS (1988) Food selection in adults and larvae of three species of Lepidoptera: a behavioural and electrophysiological study. Entomol exper et applic 49:111–121.

Boreham BW, Harvey JK (1984) Dipolar charged particles as markers for dispersion simulation in wind tunnels. J Physics E 17:994–998.

Bowden J (1982) An analysis of factors affecting catches of insects in light traps. Bull Ent Res 72:535–556.

Bowden J, Brown G, Stride T (1979) The application of X-ray spectrometry to analysis of elemental composition (chemoprinting) in the study of migration of *Noctua pronuba* L. Ecol Ent 4:199–204.

Boyan GS (1989) Is there a common "Bauplan" for insect auditory pathways? In: J Erber, R Menzel, HJ Pfluger, and D Todt (Eds), *Neural Mechanisms of Behavior*, Stuttgart: Georg Thieme, p. 73a.

Boyan GS, Fullard JH (1986) Interneurones responding to sound in the tobacco budworm moth *Heliothis virescens* (Noctuidae): morphological and physiological characteristics. J Comp Physiol A 158:391–404.

Boyan GS, Fullard JH (1988) Information processing at a central synapse suggests a noise filter in the auditory pathway of the noctuid moth. J Comp Physiol A 164:251–258.

Boyan GS, Williams L, Fullard J (1990) Organization of the auditory pathway in the thoracic ganglia of noctuid moths. J Comp Neurol 294:2–21.

Brier HB, Rogers DJ (1981) Varietal resistance to *Heliothis* species in soybeans (*Glycine max*). In: PH Twine (Ed), *Workshop on Biological Control of Heliothis spp*. Brisbane: Qld DPI Publication, pp.34–35.

Brittnacher JG, Sims SR, Ayala FJ (1978) Genetic differentiation between species of the genus *Speyeria* (Lepidoptera: Nymphalidae). Evolution 32:199–210.

Broadley RH (1981a) Parasitoids of *Heliothis* spp. larvae collected from sunflowers in south-east Queensland. In: PH Twine (Ed), *Workshop on Biological Control of Heliothis*. Brisbane: QLD DPI Publication, pp.66–68.

Broadley RH (1981b) Possible arthropod predators of *Heliothis* spp eggs and larvae in sunflowers in south-east Queensland. In: PH Twine (Ed), *Workshop on Biological Control of Heliothis*. Brisbane: QLD DPI Publication, pp.82–85.

Broadley RH (1984) Seasonal incidence and parasitism of *Heliothis* spp (Lepidoptera: Noctuidae) larvae in south Queensland sunflowers. J Aust Ent Soc 23:145–147.

Brook KD, Hearn AB (1983) Development and implementation of SIRATAC: a computer-based cotton management system. In: *Computers in Agriculture*, 1st National Conference, Proceedings. University of Western Australia.

Brooks WM (1968) Transovariol transmission of *Nosema heliothidis* in the corn earworm, *Heliothis zea* J Invertebr Pathol 11:510–517.

Brooks WM, Cranford JD (1975) Host-parasite relationships of *Nosema heliothidis*. Abstract of Papers presented at the Society for Invertebrate Pathology VIIIth Annual Meeting, Oregon State University, Corvallis.

Brown ES (1970) Nocturnal insect flight direction in relation to the wind. Proc R Ent Soc Lond (A) 45:39–43.

Brown ES, Betts E, Rainey RC (1969) Seasonal changes in distribution of the African armyworm, *Spodoptera exempta* (Wlk.) (Lep, Noctuidae), with special reference to eastern Africa. Bull Ent Res 58:661–728.

Brown LG, McClendon RW, Jones JW (1979) Computer simulation of the interaction between the cotton crop and the insect pests. Trans Amer Soc Agr Eng 22:771–774.

Browning TO (1952) The influence of temperature on the rate of development of insects, with special reference to the eggs of *Gryllus commodus*, Walker. Aust J Sci Res B 5:96–111.

Bucher GE (1960) Potential bacterial pathogens of insects and their characteristics. J Insect Pathol 2:172–195.

Burleigh JG, Farmer JH (1978) Dynamics of *Heliothis* spp larval parasitism in South-east Arkansas. Env Ent 7:692–694.

Bush GL (1966) The taxonomy, cytology, and evolution of the genus *Rhagoletis* in North America. Bull Mus Comp Zool 134:431–462.

Bush GL (1969) Sympatric host race formation and speciation in frugivorous fruit-flies of the genus *Rhagoletis* (Diptera: Tephritidae). Evolution 23:237–251.

Busvine JR (1971) *A Critical Review of the Techniques for Testing Insecticides*. 2nd Ed. Slough, England: Commonwealth Agricultural Bureau.

Butler E (1970) General random number generator. Comm Assoc Comp Mach 13:49–52.

Butler GD (1976) Bollworm: development in relation to temperature and larval food. Env Ent 5:520–522.

Butler GD, Hamilton AG (1976) Development time of *Heliothis virescens* in relation to constant temperature. Env Ent 5:759–760.

Butler GD, Hamilton AG, Proshold FI (1979) Development times of *Heliothis vires-cens* and *H. subflexa* in relation to constant temperature. Ann Ent Soc Am 72: 263–266.

Butler GD Jr, Wilson LT, Henneberry TJ (1985) *Heliothis virescens* (Lepidoptera: Noctuidae): Initiation of summer diapause. J Econ Entomol 78:320–324.

Byerly KF, Gutierrez AP, Jones RE, Luck RF (1978) A comparison of sampling methods for some arthropod populations in cotton. Hilgardia 46:257–282.

Caccone A (1985) Gene flow in cave arthropods: a qualitative and quantitive ap-proach. Evolution 39:1223–1235.

Cahill M, Easton C, Forrester N, Goodyer G (1984) Larval identification of *He-liothis punctigera* and *H. armigera* Proc. of the 1984 Australian Cotton Re-search Conference, Toowoomba, pp. 216–221.

Callahan PS (1957) Oviposition response of the corn earworm, *Heliothis zea* (Bod-die), to various wavelengths of light. Annals Ent Soc Am 50:444–452.

Callahan PS, Sparks AN, Snow JW, Copeland WW (1972) Corn earworm moth: Vertical distribution in nocturnal flight. Env Ent 1:467–503.

Campbell A, Frazier BD, Gilbert N, Gutierrez AP, Mackauer M (1974) Temperature requirements of some aphids and their parasites. J Appl Ecol 11:665–668.

Campion DG, Bettany DW, McGinnigle JB, Taylor LR (1977) The distribution and migration of *Spodoptera littoralis* (Boisduval) (Lepidoptera: Noctuidae), in re-lation to meteorology on Cyprus, interpreted from maps of pheromone trap samples. Bull Ent Res 7:193–196.

Cantello WW, Smith JS, Baumhover AH, Stanley JM, Henneberry TJ (1972) Sup-pression of an isolated population of tobacco hornworm with blacklight traps unbaited or baited with virgin female moths. Env Ent 1:253–258.

Carde RT (1984) Chemo-orientation in flying insects. In: WS Bell, RT Carde (Eds), *Chemical Ecology of Insects*, London: Chapman and Hall, pp. 111–126.

Carde RT, Hagaman TE (1979) Behavioural responses of the gypsy moth in a wind-tunnel to air-borne enantiomers of disparlure. Env Ent 8:475–484.

Carde RT, Roelofs WL, Harrison RG, Vawter AT, Brussard PF, Matuura A, Munroe E (1978) European corn borer: pheromone polymorphism or sibling species? Science 199:555–556.

Caron RE, Bradley JR Jr, Pleasants RH, Rabb RL, Stinner RE (1978) Overwinter survival of *Heliothis zea* produced on late-planted field corn in North Carolina. Env Ent 7:193–196.

Carson HL (1957) The species as a field for gene recombination. In: E Mayr (Ed), *The Species Problem*. Washington: AAAS Publication No. 50, pp. 23–38.

Cavalli-Sforza LL, Bodmer WF (1971) *The Genetics of Human Populations*. San Francisco: Freeman.

Chapman RF (1974) The chemical inhibition of feeding by phytophagous insects: A review. Bull Ent Res 64:339–363

Chapman RF (1982) *The Insects: Structure and Function*, 3rd Ed. Cambridge, Mass: Harvard University Press.

Chen GT, Graves JB (1970) Spermatogenesis of the tobacco budworm. Ann Ent Soc Am 63:1095–1104.

Clark WC (1979) Spatial structure relationship in a forest insect system: simulation models and analysis. Bull Soc Ent Suisse 52:235–257.

Close RC, Moar NT, Tomlinson AI, Lowe AD (1978) Aerial dispersal of biological material from Australia to New Zealand. Int J Biometeor 22:1-19.

Coaker TH (1959) Investigations on *Heliothis armigera* (Hb) in Uganda. Bull Ent Res 50:487-506.

Cochran WG, Cox GM (1957) *Experimental Designs*. New York: John Wiley & Sons.

Coluzzi M, Bullini L (1971) Enzyme variants as markers in the study of pre-copulatory isolating mechanisms. Nature 231:455-456.

Comins HN (1977) The development of insecticide resistance in the presence of migration. J Theor Biol 64:177-197.

Common IFB (1953) The Australian species of *Heliothis* (Lepidoptera: Noctuidae) and their pest status. Aust J Zool 1:319-344.

Common IFB (1959) A transparent light trap for the field collection of lepidoptera. J Lepid Soc 13:57-61.

Common IFB (1985) A new Australian species of *Heliothis* Ochsenheimer (Lepidoptera: Noctuidae). J Aust Ent Soc 24:129-133.

Common IFB (1990) *Moths of Australia*. Melbourne: Melbourne University Press.

Connell JH (1983) On the prevalence and relative importance of interspecific competition: evidence from field experiments. Am Nat 122:661-696.

Crow JF, Kimura M (1970) *An Introduction to Population Genetics Theory*. New York: Harper and Row.

Cullen J M (1969) The reproduction and survival of *Heliothis punctigera* Wallengren in South Australia. PhD Thesis, University of Adelaide, Australia.

Cullen JM, Browning TO (1978) The influence of photoperiod and temperature on the induction of diapause in pupae of *Heliothis punctigera*. J Insect Physiol 24: 595-602.

Cunningham RB, Lewis T, Wilson AGL (1981) Biothermal development: A model for predicting the distribution of emergence times of diapausing *Heliothis armigera*. J R Stat Soc Ser C 30:132-140.

Dallwitz MJ (1984) The influence of constant and fluctuating temperatures on development rate and survival of pupae of the Australian sheep blowfly *Lucilia cuprina*. Ent Exp et Appl 36:89-95.

Dallwitz MJ, Higgens JP (1978) User's guide to DEVAR. A computer program for estimating development rate as a function of temperature. Rep Div Ent CSIRO No. 2.

Daly JC (1989) The use of electrophoretic data in a study of gene flow in the pest species *Heliothis armigera* (Hübner) and *H. punctigera* Wallengren (Lepidoptera: Noctuidae) In: HD Loxdale, J Den Hollander (Eds) *Electrophoretic Studies on Agricultural Pests*. Systematics Special Volume No. 39. Oxford: Clarendon Press, pp. 115-141.

Daly JC, Fisk JH, Forrester NW (1988) Selective mortality in field trials between strains of *Heliothis armigera* (Hübner) (Lepidoptera: Noctuidae) resistant and susceptible to synthetic pyrethroids: functional dominance of resistance and age- class. J Econ Entomol 81:1000-1007.

Daly JC, Fitt GP (1990) Resistance frequencies in overwintering pupae and the first spring generation of *Helicoverpa armigera* (Lepidoptera: Noctuidae): Selective mortality and immigration. J Econ Entomol 83:(in press).

Daly JC, Gregg P (1985) Genetic variation in *Heliothis* in Australia: species identification and gene flow in the two pest species *H. armigera* (Hübner) and *H. punctigera* Wallengren (Lepidoptera:Noctuidae). Bull Ent Res 75:169–184.

Daly JC, Wilkenson P, Shaw DD (1981) Reproductive isolation in relation to allozymic and chromosomal differentiation in the grasshopper *Caledia captiva*. Evolution 35:1164–1179.

Danthanarayana W (1986) Lunar periodicity of insect flight and migration. In: W. Danthanarayana (Ed), *Insect Flight*. Berlin: Springer-Verlag, pp. 88–119.

David CT, Kennedy JS, Ludlow AR (1982) Finding of a sex pheromone source by gypsy moths released in the field. Nature 303:804–806.

David CT, Kennedy JS, Ludlow AR, Perry JN, Wall C (1983) A reappraisal of insect flight towards a distant point source of wind-borne odor. J Chem Ecol 8:1207–1215.

Davidson G (1956) Insecticide resistance in *Anopheles gambiae* Giles: a case of simple Medelian inheritance. Nature 178:863–864.

Davidson G (1958) Studies on insecticide resistance — resistance in *Anopheles* mosquitoes. Bull WHO 16:579–621.

Davidson J (1942) On the speed of development of insect eggs at constant temperatures. Aust J Exp Biol Med Sci 20:233–239.

Davis EE (1984) Regulation of sensitivity in the peripheral chemoreceptor systems for host-seeking behaviour by a haemolymph-borne factor in *Aedes aegypti*. J Insect Physiol 30:179–183.

Davis MTB, Miller SG (1988) Expression of a testis B-tubulin in *Heliothis virescens* during spermatogenesis. Insect Biochem 18:389–394.

de Kramer JJ (1985) The electrical circuitry of an olfactory sensillum in *Antheraea polyphemus*. J Neurosci 5:2484–2493.

Dempster JP (1960) A quantitative study of the predators on the eggs and larvae of the broom beetle, *Phytodecta olivacea* Forster, using the precipitation test. J Anim Ecol 29:149–167.

Dempster JP (1983) The natural control of populations of butterflies and moths. Bio Rev 58:461–481.

Dempster JP, Pollard E (1986) Spatial heterogeneity, stochasticity and the detection of density dependence in animal populations. Oikos 46:413–416.

den Otter CJ, Schuil HA, Sander van-Osten A (1978) Reception of host plant odours and female sex pheromone in *Adoxophyes orana* (Lepidoptera:Tortricidae): electro-physiology and morphology. Ent Exp et Appl 24:370–378.

Dennis B, Kemp WP, Beckwith RC (1986) Stochastic model of insect phenology: Estimation and testing. Env Ent 15:540–546.

Dent DR, Pawar CS (1988) The influence of moonlight and weather on catches of *Helicoverpa armigera* (Hübner) (Lepidoptera:Noctuidae) in light and pheromone traps. Bul Ent Res 78:365–377.

Dethier VG (1963) *The Physiology of Insect Senses*. New York: John Wiley & Sons, Inc.

Dethier VG (1971) A surfeit of stimuli: a paucity of receptors. Am Sci 59:706–715.

Dethier VG (1976) *The Hungry Fly: A Physiological Study of the Behavior Associated with Feeding*. Cambridge, MA: Harvard University Press.

Dethier VG (1980a) Responses of some olfactory receptors of the Eastern Tent Caterpillar (*Malacosoma americanum*) to leaves. J Chem Ecol 6:213–220.

Dethier VG (1980b) Food-aversion learning in two polyphagous caterpillars *Diacrisia virginica* and *Estigmene congrua*. Physiol Ent 5:321–325.

Dethier VG (1982) Mechanisms of host plant recognition. Ent Exp et Appl 31:49–56.

Dethier VG, Crnjar RM (1982) Candidate codes in the gustatory system of caterpillars. J Gen Physiol 79:549–569.

Dethier VG, Barton-Browne L, Smith CN (1960) The designation of chemicals in terms of the responses they elicit from insects. J Econ Entomol 53:134–136.

Dickens JC, Gutman A, Payne TL, Ryker LC, Rudinski JA (1983) Antennal olfactory responsiveness of Douglas fir beetle, *Dendroctonus pseudotsugae* Hopkins (Coleoptera:Scolytidae) to pheromones and host odours. J Chem Ecol 9:1383–1395.

Dingle H (1972) Migration strategies of insects. *Science* 175:1327–1335.

Dobzhansky T (1937) *Genetics and the Origin of Species*. 1st Ed. New York: Columbia University Press.

Dobzhansky T (1951) *Genetics and the Origin of Species*. 3rd Ed. New York: Columbia University Press.

Dobzhansky T, Ayala FJ, Stebbins GL, Valentine JW (1977) *Evolution*. San Francisco: Freeman.

Downing JD, Frost EL (1972) Recent radar observations of diurnal insect behaviour. Proc 59th Ann Meet New Jersey Mosq Exterm Assoc: 114–132.

Drake VA (1982a) The CSIRO Entomological Radar : a remote- sensing instrument for insect migration research. In: LA Wisby (Ed), *Scientific Instruments in Primary Production*. Melbourne: Australian Scientific Industry Association, pp. 63–73.

Drake VA (1982b) Insects in the sea-breeze front at Canberra: A radar study. Weather 37:134–143.

Drake VA (1984) The vertical distribution of macro-insects migrating in the nocturnal boundary layer: A radar study. Boundary-Layer Meterol 28:353–374.

Drake VA (1985) Radar observations of moths migrating in a nocturnal low-level jet. Ecol Ent 10:259–265.

Drake VA, Farrow RA (1983) The nocturnal migration of the Australian plague locust, *Chortoicetes terminifera* (Walker) (Orthoptera:Acrididae): Quantitative radar observations of a series of northward flights. Bull Ent Res 73:567–585.

Drake VA, Farrow RA (1985) A radar and aerial-trapping study of an early spring migration of moths (Lepidoptera) in inland New South Wales. Aust J Ecol 10:223–235.

Drake VA, Farrow RA (1988) The influence of atmospheric structure and motions on insect migration. Ann Rev Entomol 33:183–210.

Drake VA, Farrow RA (1989) The "aerial plankton" and atmospheric convergence. Trends Ecol Evol 4:381–385.

Drake VA, Helm KF, Readshaw JL, Reid DG (1981) Insect migration across Bass Strait during spring: A radar study. Bull Ent Res 71:449–466.

Drew RAI, Hardy DE (1981) *Dacus (Bactrocera) opiliae*, a new sibling species of the dorsalis complex of fruit flies from northern Australia (Diptera:Tephritidae). J. Aust Ent Soc 20:131–137.

Eastop VF (1973) Biotypes of aphids. In: AD Lowe (Ed), *Perspectives in Aphid Biology*. Auckland, New Zealand: Bull No 2 Entomol Soc, pp.40–51.

Eger JE (1981) Factors affecting winter survival of *Heliothis virescens* (Fabricius) and *Heliothis zea* (Boddie) (Lepidoptera:Noctuidae) and subsequent development on wild spring hosts. PhD Thesis, Texas A & M University, College Station.

Eger JE Jr, Sterling WL, Hartstack AW Jr, (1983) Winter survival of *Heliothis virescens* and *Heliothis zea* (Lepidoptera:Noctuidae) in College Station, Texas. Env Ent 12:970-975.

Eubank WP, Atmar JW, Ellington JJ (1973) The significance and thermodynamics of fluctuating versus static thermal environments on *Heliothis zea* egg development rates. Env Ent 2:491-496.

Evans M (1988) Ecology and Management of selected phytophagous arthropods species in soybean crops in South East Queensland, with special reference to *Heliothis armigera* (Hübner). PhD Thesis. University of Queensland, Australia.

Ewens WJ (1977) Population genetics theory in relation to neutralist-selectionist controversy. Adv Hum Genet 8:67-134.

Farrow RA (1975) Offshore migration and the collapse of outbreaks of the Australian plague locust (*Chortoicetes terminifera* Walk.) in south-east Australia. Aust J Zool 23:569-595.

Farrow RA (1984) Detection of transoceanic migration of insects to a remote island in the Coral Sea, Willis Island. Aust J Ecol 9:253-272.

Farrow RA, Daly JC (1987) Long-range movement as an adaptive strategy in the genus *Heliothis* (Lepidoptera:Noctuidae): a review of its occurrence and detection in four pest species. Aust J Zool 35:1-24.

Farrow RA, Dowse JE (1984) Method of using kites to carry tow nets in the upper air for sampling migrating insects and its application to radar entomology. Bull Ent Res 74:87-95.

Farrow RA, McDonald G (1987) Migration strategies and outbreaks of noctuid pests in Australia. Insect Sci Appl 8:531-542.

Felsenstein J (1982) Numerical methods for inferring evolutionary trees. Quart Rev Biol 57:379-404.

Feeny P, Rosenberry L, Carter M (1983) Chemical aspects of oviposition behavior in butterflies. In: S Ahmad (Ed), *Herbivorous Insects. Host-Seeking Behavior and Mechanisms*, New York: Academic Press, pp. 27-76.

Finch S (1980) Chemical attraction of plant-feeding insects to plants. Appl Biol 5:67-143.

Finch S (1986) Assessing host-plant finding by insects. In: JR Miller, TA Miller (Eds), *Insect-Plant Interactions*. Springer Series in Experimental Entomology, New York: Springer-Verlag, pp. 23-64.

Finney DJ (1971) *Probit Analysis*. London: Cambridge University Press.

Firempong S (1987) Some factors affecting host plant selection by *Heliothis armigera* (Hübner) (Lepidoptera:Noctuidae). PhD Thesis, University of Queensland, Australia.

Firempong S, Zalucki MP (1990a) Host Plant Preferences of Populations of *Helicoverpa armigera* (Hübner) (Lepidoptera:Noctuidae) from different geographic locations. Aust J Zool 37:665-673.

Firempong S, Zalucki MP (1990b) Host Plant Selection by *Helicoverpa armigera* (Hübner) (Lepidoptera:Noctuidae): Role of certain plant attributes. Aust J Zool 37:665-683.

Fisk JH (1989) Karyotype and achiasmatic female meiosis in *Helicoverpa armigera* (Hübner) and *H. punctigera* (Wallengren) (Lepidoptera:Noctuidae). Genome 32:967-971.

Fisk JH, Daly JC (1989) Electrophoresis of *Helicoverpa armigera* (Hbner) and *H. punctigera* (Wallengren) (Lepidoptera:Noctuidae): genotype expression in eggs and allozyme variations between life stages. J Aust Ent Soc 28:191-192.

Fitt GP (1986) The influence of a shortage of hosts on the specificity of oviposition behaviour in species of *Dacus* (Diptera:Tephritidae). Physiol Ent 11:133-143.

Fitt GP (1987) Ovipositional responses of *Heliothis* females to host plant variation in cotton, *Gossypium hirsutum*. In: V Labeyrie, G Fabres, D Lachaise (Eds.) *Insect — Plant. Pro. 6th Int Symp on Insect-Plant Relationships* (Pau 1986) pp.289-294. Dr. W Junk, Dordrecht.

Fitt GP (1989) The ecology of *Heliothis* in relation to agroecosystems. Ann Rev Entomol 34:17-52

Fitt GP (1991) Host selection in the Heliothinae. In: J Ridsdill-Smith, W Bailey (Eds), *Reproductive Behaviour in Insects — Individuals and Populations*, London: Chapman and Hall (in press).

Fitt GP, Daly JC (1990) Abundance of overwintering pupae and the spring generation of *Helicoverpa* spp. (Lepidoptera:Noctuidae) in northern New South Wales, Australia: Implications for pest management. J Econ Entomol 83:(in press).

Foley DH (1981) Pupal development rate of *Heliothis armiger* (Hübner) (Lepidoptera:Noctuidae) under constant and fluctuating temperatures. J Aust Ent Soc 20:13-20.

Forrester NW (1981) Biological control of agent of *Heliothis* spp. in sunflowers. In PH Twine (Ed), *Workshop on biological control of Heliothis*. Brisbane: QLD DPI publication.

Forrester NW, Cahill M (1987) Management of insecticide resistance in *Heliothis armigera* (Hübner) in Australia. In: M Ford, BPS Khambay, DW Holloman, RM Sawicki (Eds), *Combating Resistance in Xenobiotics: Biological and Chemical Approaches*, Chichester, England: Ellis Horwood, pp.127-137.

Forrester NW (1989) Six years experience with the strategy. The Australian Cotton Grower 10(3):62-64.

Foster GG, Whitten MJ, Prout T, Gill, R. (1972). Chromosome rearrangements for the control of insect pests. Science 176:875-880.

Fox KJ (1969) Recent records of migrant Lepidoptera in Taranaki. N.Z. Ent. 4:6-1

Fox KJ (1975) Migrant Lepidoptera in New Zealand 1973-1974. N.Z. Ent. 6:66-69.

Fox LR, Morrow PA (1981) Specialisation: Species property or local phenomenon. Science 211:887-893.

Frazer BD, Gilbert NE (1976) Coccinellids and aphids: A quantitative study of the impact of adult ladybirds. J Ent Soc Br Col 73:33-56.

Frazier JL, Hanson FE (1986) Electrophysiological recording and analysis of insect chemosensory responses. In: JR Miller, TA Miller (Eds), *Insect-Plant Interactions*. Springer Series in Experimental Entomology, New York: Springer-Verlag, pp.285-330.

Frost SW (1957) The Pennsylvania insect light trap. J Econ Entomol 50:287-292.

Frost SW (1958) Insects captured in light traps with and without baffles. Can Ent 90:566-567.

Fullard JH (1984) Acoustic relationships between tympanate moths and the Hawaiian hoary bat (*Lasiurus cinereus semotus*). J Comp Physiol A 155:795–801.

Fullard JH (1987) Sensory ecology and neuroethology of moths and bats:interactions in a global perspective. In: MB Fenton, P Racey, JMV Rayner (Eds), *Recent Advances in the Study of Bats*, Cambridge: Cambridge University Press, pp. 244–272.

Futuyma DJ (1979) *Evolutionary Biology*. Massachusetts: Sinauer Associates.

Futuyma DJ, Peterson SC (1985) Genetic variation in the use of resources by insects. Ann Rev Entomol 30:217–238.

Fye RE (1971) Temperature in the plant parts of short-staple cotton. J Econ Entomol. 64:1432–1435.

Gatehouse AG (1987) Migration and low population density in armyworm (Lepidoptera:Noctuidae) life histories. Insect Sci Applic 8:573–580.

Gaugler RR, Brooks WM (1975) Sublethal effect of infection by *Nosema heliothidis* in the corn earworm, *Heliothis zea*. J Invertebr Pathol 26:57–63.

Geier PW, Clark LR (1976) On the developing use of modelling in insect ecology and pest management. Aust J Ecol 1:119–127.

Geier PW, Hillman TJ (1971) An analysis of th life system of the codling moth in south-east Australia. Proc Ecol Soc Aust 6:203–243.

Getzin LW (1961) *Spicaria rileyi* (Farlow) Charles, an entomogenous fungus of *Trichoplusia ni* (Hübner). J Insect Pathol 3:2–10.

Gilbert N, Gutierrez AP (1973) A plant-aphid-parasite relationship. J Anim Ecol 42:323–340.

Gilbert N, Gutierrez AP, Frazer BD, Jones RE (1976) *Ecological Relationships*. Reading: Freeman.

Glick PA (1965) Review of collections of Lepidoptera by airplane. J Lepid Soc 19:129–137.

Goldsmith FB, Harrison CM, Morton AJ (1986). Description and analysis of vegetation. In: PD Moore and SB Chapman (Eds), *Methods in Plant Ecology*. Oxford: Blackwell Scientific Publication, pp. 437–524.

Gonzalez D, Gordh G, Thompson SN, Adler J (1979) Biotype discrimination and its importance to biological control. In: MA Hoy and JJ McKelvey, Jr (Eds), *Genetics in Relation to Insect Management*. New York: The Rockefeller Foundation. pp. 129–136.

Goodenough JL, Snow JW (1973) Increased collection of tobacco budworm by electric grid traps as compared with blacklight and sticky traps. Econ Ent 66:450–453.

Gothilf S, Kehat M, Dunkelblum E, Jacobson M (1979) Efficacy of (Z)-11-hexadecenal and (Z)-11 tetradecenal as sex attractants for *Heliothis armigera* on two different dispensers. J Econ Entomol 72:718–720.

Graham HM, Wolfenbarger DA (1977) Tobacco budworm: Labelling with rubidium in the laboratory. J Econ Entomol 70:800–802.

Graham HM, Wolfenbarger DA, Nosky JB (1978a) Labelling plants and their insect fauna with rubidium. Env Ent 7:379–383.

Graham HM, Wolfenbarger DA, Nosky JB, Hernandez NS Jr, Llanes JR, Tamayo JA (1978b) Use of rubidium to label corn earworm and fall armyworm for dispersal studies. Env Ent 7:435–438.

Grant AJ, O'Connell RJ (1986) Neurophysiological and morphological investigations of pheromone-sensitive sensilla on the antenna of male *Trichoplusia ni*. J Insect Physiol 32:503–515.

Greenbank DO (1957) The role of climate and dispersal in the initiation of outbreaks of the spruce budworm in New Brunswick. II. The role of dispersal. Can J Zool 35:385–403.

Greenbank DO, Schaefer GW, Rainey RC (1980) Spruce budworm (Lepidoptera: Tortricidae) moth flight and dispersal: New understanding from canopy observations, radar, and aircraft. Mem Ent Soc Can 110:1–49.

Greenstone MH (1979) A line transect density index for wolf spiders (*Paradosa* spp.), and a note on the applicability of catch per unit effort methods to entomological studies. Ecol Ent 4:23–29.

Gregg PC (1981) The Adaptation of a Locust to its Physical Environment. PhD Thesis, Australian National University, Canberra, Australia.

Gregg PC (1982) A simulation model of the development of *Chortoicetes terminifera* (Orthoptera:Acrididae) under fluctuating temperatures. Proc. III Australian Conf Grassl Invert Ecol 117–125.

Gregg PC, McDonald G, Bryceson KP (1989) The occurrence of *Heliothis punctigera* Wallengren and *H. armigera* (Hübner) in inland Australia. J Aust Ent Soc 28:135–140.

Griffith IP, Smith AM, Williamson WEP (1979) Raising potato moth larvae (*Phthorimaea operculella* (Zeller) (Lepidoptera:Gelechiidae) in isolation. J. Aust Ent Soc 18:348.

Griffiths KJ, Holling CS (1969) A competition submodel for parasites and predators. Can Ent 101:785–818.

Grosberg RK, Quinn JF (1986). The genetic control and consequences of kin recognition by the larvae of a colonial marine invertebrate. Nature 322:456–459.

Gross HR (1984) *Spodoptera frugiperda* (Lepidoptera:Noctuidae): Deterrence of oviposition by aqueous homogenates of fall armyworm and corn earworm larvae applied to whorl-stage corn. Env Ent 13:1498–1501.

Gross HR, Carpenter JE, Sparks AN (1983) Visual acuity of *Heliothis zea* (Lepidoptera:Noctuidae) males as a factor influencing the efficiency of pheromone traps. Env Ent 12:844–847.

Guerrin PM, Visser JH (1980) Electroantennogram responses of the carrot fly, *Psila rosae*, to volatile plant components. Physiol Ent 5:111–119.

Guerrin PM, Stadler E (1982) Host odour perception in three phytophagous Diptera — a comparative study. In: JH Visser, AK Minks (Eds), *Proc. 5th International Symposium on Insect-Plant Relationships*, Pudoc 1982, Wageningen, pp. 95–105.

Gunning RV (1988) Pyrethroid resistance in unsprayed *Heliothis*. The Aust Cotton Grower 9(1):26–28.

Gunning RV, Easton CS (1989) Pyrethroid resistance in *Heliothis armigera* (Hübner) collected from unsprayed crops in New South Wales 1983–1987. J Aust Ent Soc 28:57–61.

Gunning RV, Easton CS, Greenup LR, Edge VE (1984) Pyrethroid resistance in *Heliothis armiger* (Hübner) (Lepidoptera:Noctuidae) in Australia. J Econ Entomol 77:1283–1287.

Gutierrez AP, Havenstein DE, Nix HA, Moore PA (1974) The ecology of *Aphis craccivora* Koch and subterranean clover stunt virus in south-east Australia. II. A model of cowpea aphid in temperate pastures. J Appl Ecol 11:1–20.

Hackett DS, Gatehouse AG (1982a) Diapause in *Heliothis armigera* (Hübner) and *H. fletcheri* (Hardwick) (Lepidoptera:Noctuidae) in the Sudan Gezira. Bull Ent Res 72:409–422.

Hackett DS, Gatehouse AG (1982b) Studies on the biology of *Heliothis* spp in Sudan. In: W Reed, V Kumble (Eds), *Proceedings of the International Workshop on Heliothis Management*. 15–20 November 1981, ICRISAT Center, Patancheru, A.P., India.

Haggis MJ (1982) Distribution of *Heliothis armigera* eggs on cotton in the Sudan Gezira:spatial and temporal changes and their possible relation to weather. In: W Reed,V Kumble (Eds), *Proceedings of the International Workshop on Heliothis Management*. 15–20 November 1981, ICRISAT Center, Patancheru, A.P., India.

Haile DG, Snow, JW, Young JR (1975) Movement by adult *Heliothis* released on St Croix to other islands. Env Ent 4:225–226.

Hanson FE (1976) Comparative studies on induction of food choice preferences in Lepidopterous larvae. Symp Biol Hungary 16:71–77.

Hanson FE (1983) The behavioral and neurophysiological basis of food-plant selection by lepidopterous larvae. In: S Ahmad (Ed), *Herbivorous Insects. Host-Seeking Behavior and Mechanisms*, New York: Academic Press, pp. 3–26.

Harcourt DG, Yee JM (1982) Polynomial algorithm for predicting the duration of insect life stages. Env Ent 11:581–584.

Hardwick DF (1965) The Corn Earworm Complex. Mem Entomol Soc Can 40:1–247.

Harris H, Hopkinson DA (1976) *Handbook of Enzyme Electrophoresis on Human Genetics*. Amesterdam: North Holland Publ. Co.

Harrison RG, Vawter AT (1977) Allozyme differentiation between pheromone strains of the European corn borer, *Ostrinia nubilalis*. Ann Ent Soc Am 70:717–720.

Harrison RG, Rand DM, Wheeler WC (1987) Mitochondrial DNA variation in field crickets across a narrow hybrid zone. Mol Biol Evol 4:144–158.

Hartstack AW (1982) Modelling and forecasting *Heliothis* populations. In: W Reed, V Kumble (Eds), *Proceedings of the International Workshop on Heliothis Management*. 15–20 November, 1981, ICRISAT Center, Patancheru, A.P., India.

Hartstack AW, Hollingsworth JP (1974) A computer model for predicting *Heliothis* populations. Trans Amer Soc Agric Eng 17:112–115.

Hartstack AW, Witz JA (1981) Estimating field populations of tobacco budworm moths from pheromone trap catches. Env Ent 10:908–914.

Hartstack AW, Witz JA, Buck DR (1979a) Moth traps for the tobacco budworm. J Econ Entomol 72:519–522.

Hartstack AW, Hollingsworth JP, Ridgeway RL, Hunt HH (1971) Determination of trap spacings required to control an insect population. J Econ Entomol 64:1090–1100.

Hartstack AW, Witz JA, Hollingsworth JP, Ridgeway RL, Lopez JD (1976) MOTHZV-2: A computer simulation of *Heliothis zea* and *Heliothis virescens* population dynamics. User manual. USDA. ARS-S-127.

Hartstack AW, Hollingsworth JP, Witz JA, Buck DR, Lopez JD, Hendricks DE

(1978) Relation of tobacco budworm catches in pheromone-baited traps to field populations. Southw Entomol 3:43–51.

Hartstack AW, Lopez JD, Mueller RA, Witz JA (1986) Early season occurrence of *Heliothis* spp. in 1982: evidence of long range migration of *Heliothis zea*. In: AN Sparks (Ed), *Long-Range Migration of Moths of Agronomic Importance to the United States and Canada: Specific Examples of Occurrence and Synoptic Weather Patterns Conducive to Migration*. U.S.D.A. ARS-43, pp.48–68.

Hartstack AW, Lopez JD, Mueller RA, Sterling WL, King EG, Witz JA, Eversull AC (1982) Evidence of long range migration of *Heliothis zea* (Boddie) into Texas and Arkansas. Southw Entomol 7:188–201.

Hartstack AW, Hollingsworth JP, Ridgeway RL, Coppedge JR (1973) A population dynamics study of the bollworm and tobacco budworm with light traps. Env Ent 2:244–252.

Hassell MP (1985) Insect natural enemies as regulating factors. J Anim Ecol 54:323–334.

Hassell MP (1987) Detecting regulation in patchily distributed animal populations. J Anim Ecol 56:705–713.

Hassell MP, Southwood TRE, Reader PM (1987) The dynamics of the vibernum whitefly (*Aleurotrachelus jelinekii*): A case study of population regulation. J Anim Ecol 56:283–300.

Haynes KF, Baker TC (1989) An analysis of anemotactic flight in female moths stimulated by host odour and comparison with males response to sex pheromone. Physiol Ent 14:279–289.

Heard TA (1985) Induced resistance of cotton to feeding by *Heliothis armigera* (Lepidoptera:Noctuidae). M.Agr. St. Thesis: University of Queensland, Australia.

Hearn AB, Brook KD (1983) SIRATAC—A case study in pest management of cotton. In: DE Byth, MA Foale, VE Mungomery, SE Wallis (Eds), *New Technology for Field Crop Production*. Brisbane: Australian Institute of Agricultural Science, pp.199–211.

Hearn AB, da Roza GD (1985) A simple model for crop management applications for cotton (*Gossypium hirsutum* L.). Field Crop Res 12:49–69.

Hearn AB, Ives PM, Room PM, Thompson NJ, Wilson LT (1981) Computer-based cotton pest management in Australia. Field Crop Res 4:321–332.

Hendricks DE, Graham HM (1970) Oil soluble dye in larval diet for tagging moths, eggs and spermatophores of tobacco budworms. J Econ Entomol 63:1019–1020.

Hendricks DE, Hollingsworth JP, Hartstack AW (1972) Catch of tobacco budworm moths influenced by colour of sex lure traps. Env Ent 1:48–51.

Hendricks DE, Graham HM, Guerra RJ, Perez CT (1973a) Comparison of the numbers of tobacco budworms and bollworms caught in sex pheromone traps vs. blacklight traps in lower Rio Grande valley, Texas. Env Ent 2:911–914.

Hendricks DE, Graham HM, Raulston JR (1973b) Dispersal of sterile tobacco budworms from release points in northeastern Mexico and southern Texas. Env Ent 2:1085–1088.

Hendricks DE, Lingren PD, Hollingsworth JP (1975) Numbers of bollworms, tobacco budworms and cotton leafworms caught in traps equipped with fluorescent lamps of five colours. J Econ Entomol 68:645–649.

Hendricks DE, Hartstack AW, Shaver TN (1977) Effect of formulations and dispensers on attractiveness of virelure to the tobacco budworm. J Chem Ecol 3:497–506.

Hendricks DE, Perez CT, Guerra RJ (1980) Effects of nocturnal wind on performance of two sex pheromone traps for noctuid moths. Env Ent 9:483–485.

Hendrix WH III, Mueller TF, Phillips JR, Davis OK (1987) Pollen as an indicator of long-distance movement of *Heliothis zea* (Lepidoptera:Noctuidae). Env Ent 16:1148–51

Higley LG, Pedigo LP, Ostlie KR (1986) DEGDAY: a program for calculating degree-days, and assumptions behind the degree-day approach. Env Ent 15:999–1016.

Hillhouse TL, Pitre HN (1976) Oviposition by *Heliothis* on soybeans and cotton. J Econ Entomol 69:144–146.

Hoffman JD, Ertle LR, Brown JB, Lawson FR (1970) Techniques for collecting,holding and determining parasitism of Lepidopteran eggs. J Econ Entomol 63:1367.

Hogg DB, Nordheim EV (1983) Age-specific survivorship analysis of *Heliothis* spp. populations on cotton. Res Popul Ecol 25:280–297.

Hogsette JA (1983) An attractant self-marking device for marking field populations of stable flies with fluorescent dusts. J Econ Entomol 76:510–514.

Holling CS (1964) The analysis of complex population processes. Can Ent 96:335–347.

Holling CS (1966) The functional response of invertebrate predators to prey density. Mem Ent Soc Can 48:1–86.

Hollingsworth JP, Hartstack AW, Buck DR, Hendricks DE (1978) Electric and nonelectric moth traps baited with synthetic sex pheromone of the tobacco budworm. USDA, ARS-s-173.

Holloway JD (1977) *The Lepidoptera of Norfolk Island: Their Biogeography and Ecology*. The Hague: Junk.

Holstein MH (1957) Cytogenetics of *Anopheles gambiae*. Bull WHO 16:456–458.

Horn HS (1983) Some theories about dispersal. In: IR Swingland, PJ Greenwood (Eds), *The Ecology of Animal Movement*, Oxford: Oxford University Press,pp. 54–62.

Howe RW (1967) Temperature effects on embryonic development in insects. Ann Rev Entomol 12:15–42.

Howell HN Jr, Granovsky TA (1982) An infrared viewing system for studying insect behaviour. Southw Entomol 7:36–38.

Hsia T-s, Tsai S-m, Ten H-s (1963) Studies of the regularity of outbreak of the oriental armyworm *Pseudaletia separata* Walker II. Observations on migratory activity of the moths across the Chili Gulf and Yellow Sea of China. (In Chinese, with English summary) Acta Entomol Sin 12:552–564. (Rev Appl Ent (A) 52:353–354.)

Hsiao HS (1972) *Attraction of Moths to Light and to Infrared Radiation*. San Francisco: San Francisco Press.

Huffaker CB (1958) Experimental studies on predation: Dispersion factors and predator-prey oscillations. Hilgardia 27:343–383.

Hughes RD (1979) Movement in population dynamics. In: RL Rabb, GG Kennedy (Eds), *Movement of Highly Mobile Insects: Concept and Methodology in Research*. Raleigh, North Carolina: University Graphics, pp.14–34.

Hughes RD, Gilbert N (1968) A model of an aphid population — A general statement. J Anim Ecol 37:553–564.

Hughes RD, Nicholas WL (1974) The spring migration of the bushfly (*Musca vetustissima* Walk.):evidence of displacement provided by natural population markers including parasitism. J Anim Ecol 43:411–428.

Hughes RD, Sands P (1979) Modelling bushfly populations. J Appl Ecol 16:117–139.

Hull CH, Nie NH (1979) *SPSS Update: New Procedures and Facilities for Releases 7 and 8*. Sydney: McGraw-Hill.

Hummel HE, Miller TA (Eds) *Techniques in Pheromone Research*. Springer Series in Experimental Entomology, Springer-Verlag, New York.

Hurlbert SH (1984) Pseudoreplication and the design of ecological field experiments. Ecol Mon 54:187–211.

Ignoffo CM (1965) The nuclear-polyhedrosis virus of *Heliothis zea* (Boddie) and *Heliothis virescens* (Fabricius). II. Propagation and biology of diet-reared *Heliothis*. J Invertebr Pathol 7:217–226.

Ignoffo CM (1971) Microbial insecticides: No — yes: Now — when! Proc Tall Timbers Conf Ecol Anim Control by Habitat Management 2:41–57.

Ignoffo CM (1973) Development of a viral insecticide:concept to commercialization. Expl Parasit 33:380–406.

Ignoffo CM, Marston NL, Hotstetter DL, Puttler B, Bell JV (1976) Natural and induced epizootics of *Nomuraea rileyi* in soybean caterpillars. J Invertebr Pathol 27:191–198.

Ives PM, Wilson LT, Cull PT, Palmer WA, Haywood C, Thomson N, Hearn AB, Wilson AGL (1984) Field use of SIRATAC: an Australian computer-based pest management system for cotton. Prot Ecol 6:1–21.

Jackson DM, Severnson RF, Johnson AW, Chaplin JF, Stephenson MG (1984) Ovipositional response of tobacco budworm moths (Lepidoptera:Noctuidae) to cuticular chemical isolates from green tobacco leaves. Env Ent 13:1023–1039.

Jackson RC (1971) The karyotype in systematics. Ann Rev Ecol Syst 2:327–368.

Jefferson RN, Rubin RE, McFarland SU, Shorey HH (1970). Sex pheromones of noctuid moths. XXII. The external morphology of the antennae of *Trichoplusia ni*, *Heliothis zea*, *Prodenia ornighogalli* and *Spodoptera exigua*. Ann Ent Soc Am 63:1227–1238.

Jermy T, Hansen FE, Dethier VG (1968) Induction of specific food preference in lepidopterous larvae. Ent Exp et Appl 11:211–230.

Jermy T, Bernays EA, Szentezi A (1982) The effect of repeated exposure to feeding deterrents on their acceptability to phytophagous insects. In: JH Visser, AK Minks (Eds), *Proceedings 5th International Symposium on Insect-Plant Relationships*, Wageningen. Pudoc 1982, Wageningen, pp. 25–32 ,

Johnson AW (1979) Tobacco budworm damage to flue-cured tobacco at different plant growth stages. Env Ent 72:602–605.

Johnson CG (1960) A basis for a general system of insect migration and dispersal by flight. Nature 186:348–350.

Johnson DR (1983) Relationship between tobacco budworm (Lepidoptera:Noctuidae) catches when using pheromone traps and egg counts in cotton. J Econ Entomol 76:182–183.

Johnson MW, Stinner RE, Rabb RL (1975) Ovipositional response of *Heliothis zea* (Boddie) to its major hosts in North Carolina. Env Ent 4:291–297.

Johnson WE, Carson HL, Kaneshiro KY, Steiner WWM (1975) Genetic variation in Hawaiian *Drosophila* II. Allozymic differentiation in the *D. planitibia* subgroup. In: CL Markert (ED), *Isozymes IV, Genetics and Evolution*. New York: Academic Press, pp.563–584.

Johnston SJ (1985) Low level augmentation of *Trichogramma pretiosum* and naturally occurring *Trichogrammaga* spp parasitism of *Heliothis* in cotton in Louisiana. Env Ent 14:28–31.

Jones JW, Brown LG, Hesketh JD (1980) COTCROP:a computer model for cotton growth and yield. In: JD Hesketh, JW Jones (Eds), *Predicting Photosynthesis in Ecosystem Models*. USA:CRC Press, West Palm Beach, Florida, pp.209–241.

Jones RE (1977) Movement patterns and egg distribution in cabbage butterflies. J Anim Ecol 46:195–212.

Jones RE (1987) Ants, parasitiods, and the cabbage butterfly *Pieris rapae*. J Anim Ecol 56:739–749.

Jones RE, Nealis VG, Ives PM, Scheermeyer E (1987) Seasonal and spatial variation in juvenile survival of the cabbage butterfly *Pieris rapae*:Evidence for patchy density-dependence. J Anim Ecol 56:723–737.

Joyce RJV (1982a) A critical review of the role of chemical pesticides in *Heliothis* management.In: W Reed W, V Kumble (Eds), *Proceedings of the International Workshop on Heliothis Management*. 15–20 November, 1989, ICRISAT Center, Patancheru, A.P., India.

Joyce RJV (1982b) The control of migrant pests. In: DJ Aidley (Ed), *Animal Migration*. Cambridge: Cambridge University Press, pp. 209–229.

Kaissling KE (1974) Sensory transduction in insect olfactory receptors. In: L Jaenicke (Ed), *Biochemistry of Sensory Function*, New York: Springer-Verlag, pp. 244–273.

Kaissling KE, Meng LZ, Bestmann HJ (1989) responses of bombykol receptor cells to (Z,E)-4,6-hexadecadiene and linolool. J Comp Physiol A 165:147–154.

Kammer AE (1971) The motor output during turning flight in a hawkmoth, *Manduca sexta*. J Insect Physiol 17:1073–1086

Karandinos MG (1976) Optimal sample size and comments on some published formulae. Bull Ent Soc Am 22:193–196.

Kay IR (1981) The effect of constant temperatures on the development time of eggs of *Heliothis armigera* (Hübner) (Lepidoptera:Noctuidae). J Aust Ent Soc 20:155–156.

Kay IR (1982a) The incidence and duration of pupal diapause in *Heliothis armiger* (Hübner) (Lepidoptera:Noctuidae) in southeast Queensland. J Aust Ent Soc 21:263–266.

Kay IR (1982b) Overwintering by three parasites of *Heliothis armiger* (Lepidoptera:Noctuidae) in southeast Queensland. J Aust Ent Soc 21:267–268.

Kay IR, Noble RM, Twine PH (1979) The effect of gossypol in artificial diet on the growth and development of *Heliothis punctigera* and *Heliothis armigera* (Lepidoptera:Noctuidae). J Aust Ent Soc 18:229–232.

Kehat M, Greenberg S (1978) Efficiency of the synthetic sex attractant and the effect of trap size on captures of *Spodoptera littoralis* males in water traps and dry funnel traps. Phytoparasitica 6:79–83.

Keil TA (1989) Fine structure of the pheromone sensitive sensilla on the antenna of the hawkmoth, *Manduca sexta*. Tissue Cell 21:139–151.

Kelker GH (1940) Estimating deer populations by a differential hunting loss in the sexes. Proc Utah Ac Sci Arts Letters 17:65–69.

Kennedy JS (1965) Mechanisms of host plant selection. Ann Appl Biol 56:317–322

Kennedy JS (1977) Behaviourally discriminating assays of attractants and repellents. In: HH Shorey, JJ McKelvey (Eds), *Chemical Control of Insect Behavior. Theory and Application*, New York: Wiley-Interscience, pp. 215–230.

Kennedy JS (1986) Some current issues in orientation to odour sources. In: TL Payne, MC Birch, CEJ Kennedy (Eds), *Mechanisms in Insect Olfaction*, Oxford: Clarendon Press, pp. 11–25.

Kennedy JS, Marsh D (1974) Pheromone-regulated anemotaxis in flying moths. Science 184:999–1001.

Kennedy JS, Way MJ (1979) Summing up the conference. In: RL Rabb, GG Kennedy (Eds), *Movement of Highly Mobile Insects: Concepts and Methodology in Research*. Raleigh, North Carolina: University Graphics, pp. 446–456.

Key KHL (1976) A generic and suprageneric classification of the Morabinae (Orthoptera:Eumasticadae), with descriptions of the type species and a bibliography of the subfamily. Aust J Zool Suppl Ser No. 37:1–185.

Kirkpatrick TH (1961) Comparative morphological studies of *Heliothis* species (Lepidoptera:Noctuidae) in Queensland. Queensl J Agric Sci 18:179–194.

Kirkpatrick TH (1962) Notes on the life-histories of species of *Heliothis* (Lepidoptera: Noctuidae) from Queensland. Queensl J Agric Sci 19:567–570.

Kitching RL (1977) Time, resources and population dynamics in insects. Aust J Ecol 2:31–42.

Kitching RL (1983) *Systems Ecology: An Introduction to Ecological Modelling*. St Lucia: University of Queensland Press, Australia.

Kitching RL, Zalucki MP (1982) Component analysis and modelling of the movement process: Analysis of simple tracks. Res Popul Ecol 24:224–238.

Klijnstra JW (1982) Perception of the oviposition deterrent pheromone in *Pieris brassicae*. In: JH Visser, AK Minks (Eds), *Proceedings 5th International Symposium on Insect-Plant Relationships*, Wageningen. Pudoc 1982, Wageningen, pp. 145–152.

Kogan M (1977) The role of chemical factors in insect/plant relationships. *Proceedings 15th International Congress of Entomology*, 1976, Washington, pp 211–227.

Kogan M (1986) Bioassays for measuring quality of insect food. In: JR Miller, TA Miller TA (Eds), *Insect-Plant Interactions*. Springer Series in Experimental Entomology, New York: Springer-Verlag, pp. 155–189.

Knipling EF (1979) *The Basic Principles of Insect Population Suppression and Management*. Washington: Agricultural Handbook No. 512. USDA.

Knipling EF, Stadelbacher EA (1983) The rationale for areawide management of *Heliothis* (Lepidoptera:Noctuidae) populations. Bull Ent Soc Am 29:29–37.

Komarova OS (1959) On the condition determining diapause of hibernating pupae of *Chloridea obsoleta* F. (Lepidoptera:Noctuidae). Ent Rev Wash 38:318–325.

Kondoh Y, Obara Y (1982). Anatomy of motoneurones innervating mesothoracic indirect flight muscles in the silkmoth, *Bombyx mori*. J Exp Biol 98:23–37.

Kramer JP (1959). On *Nosema heliothidis* Lutz and Splendore, a microsporidian parasite of *Heliothis zea* (Boddie) and *Heliothis virescens* (Fabricius) (Lepidoptera, Phalaenidae). J Insect Pathol 1:297–303.

Kreitman M, Aguade M (1986) Genetic uniformity in two populations of *Drosophila melanogaster* as revealed by filter hybridization of four-nucleotide-recognizing restriction enzyme digests. Proc Natl Acad Sci (USA) 83:3562–3566.

Kretzschmar GP (1948) Soybean insects in Minnesota with special reference to sampling techniques. J Econ Entomol 41:586–591.

Lam JJ, Baumhover AH (1982) Nocturnal response of *Heliothis virescens* (Lepidoptera:Noctuidae) to artificial light and sex pheromones. Env Ent 11:1032–1035.

Lambert DM, Paterson HEH (1982) Morphological resemblance and its relationship to genetic distance measures. Evol Theor 5:291–300.

Lansman RA, Shade RO, Shapira JF, Avise JC (1981) The use of restriction endonucleases to measure mitochondrial DNA sequence relatedness in natural populations. III. Techniques and potential applications. J Mol Evol 17:214–226.

Laster ML (1972) Interspecific hybridization of *Heliothis virescens* and *H. subflexa*. Env Ent 1:682–687.

Laster ML, Goodpasture CE, King EG, Twine P (1985) Results from crossing the bollworms *Helicoverpa armigera* and *H. zea* in search of backcross sterility. *Proceedings of the Beltwide Cotton Conference*, New Orleans 1985. pp. 146–147.

Lateef SS (1985) Gram pod borer (*Heliothis armigera*) resistance in chickpeas. Agricul Ecosyst and Env 14:95–102.

Legg BJ, Strange SA, Wall C, Perry JN (1980) Diffusion of gaseous plumes within crop canopies. Rothamsted Experiment Station Report 1979, Part 1, p. 162.

Leslie PH (1945) On the use of matrices in certain population mathematics. Biometrika 33:183–212.

Lewis AC, van Emden HF (1986) Assays for insect feeding. In: JR Miller, TA Miller (Eds), *Insect-Plant Interactions*. Springer Series in Experimental Entomology, New York: Springer-Verlag, pp. 95–120.

Lewis WJ, Brazzel JR (1968) A three year study of parasites of the bollworm and the tobacco budworm in Mississippi. J Econ Entomol 61:673–676.

Lewis WJ, Sparks AN, Jones RL, Barras DJ (1972) Efficiency of *Cardiochiles nigriceps* as a parasites of *Heliothis virescens* on cotton. Env Ent 1:468–471.

Lewontin RC (1974) The Genetic Basis of Evolutionary Change. New York: Columbia University Press.

Lewontin RC (1985) Population genetics. Ann Rev Genet 19:81–102.

Lin C-s, Sun C-j, Chen R-l, Chang JT-p (1963) Studies on the regularity of the outbreak of the oriental armyworm, *Leucania separata* Walker. I. The early spring migration of the oriental armyworm moths and its relation to winds. (In Chinese, with English summary.) Acata Entomol Sin 12:243–261. (Rev Appl Ent (A) 52:193.)

Lingren PD, Wolf W (1982) Nocturnal activity of the tobacco budworm and other insects. In: JL Hatfield, IJ Thomason (Eds), *Biometeorology in Integrated Pest Management*, New York: Academic Press, pp. 211–228.

Lingren PD, Greene GL, Davis DR, Baumhover AH, Henneberry TJ (1977) Nocturnal behaviour of four Lepidopteran pests that attack tobacco and other crops. Ann Ent Soc Am 70:161–167.

Lingren PD, Lukefahr MJ, Diaz M, Hartstack AW (1978a) Tobacco budworm control in caged cotton with a resistent variety, augmentative releases of *Campoletis sonorenus* and natural control by other beneficial species. J Econ Entomol 71:739–745.

Lingren PD, Sparks AN, Raulston JR, Wolf WN (1978b) Applications for nocturnal studies of insects. Bull Ent Soc Am 24:206–212.

Lingren PD, Sparks AN, Raulston JR (1982) The potential contribution of moth behaviour research to *Heliothis* management. In: W Reed, V Kumble (Eds), *Proceedings of the International Workshop on Heliothis Management*, 15–20 November, 1981, ICRISAT Center, Patancheru, A.P., India.

Linn CE, Campbell MG, Roelofs WL (1986) Male moth sensitivity to multicomponent pheromones:critical role of female-released blend in determining the functional role of components and active space of the pheromone. J Chem Ecol 12:659–668.

Logan JA, Wollkind, DJ, Hoyt SC, Tanigoshi LK (1976) An analytical model for description of temperature dependent rate phenomena in arthropods. Env Ent 5:1133–1140.

Logan JA, Stinner RE, Rabb RL, Bacheler JS (1979) A descriptive model for predicting spring emergence of *Heliothis zea* populations in North Carolina. Env Ent 8:141–146.

Lopez JD (1986) Thermal requirements for diapause in laboratory cultures of *Heliothis zea* and *H. virescens* (Lepidoptera:Noctuidae). Env Ent 15:919–923.

Lopez JD, Hartstack AW (1985) Comparison of diapause development in *Heliothis zea* and *H. virescens* (Lepidoptera:Noctuidae). Ann Ent Soc Am 78:415–422.

Lopez JD, Hartstack AW, Beach R (1984) Comparative pattern of emergence of *Heliothis zea* and *H. virescens* (Lepidoptera:Noctuidae) from overwintering pupae. J Econ Entomol 76:1421–1426.

Lopez JD, Hartstack AW, Witz JA, Hollingsworth JP (1979) Relationship between bollworm oviposition and moth catches in blacklight traps. Env Ent 8:42–45.

Loughton BG, Derry C, West AS (1963) Spiders and the spruce budworm. Mem Ent Soc Can 31:249–268.

Loukas M, Delidakis C, Kafatos FC (1986) Genomic blot hybridization as a tool of phylogenetic analysis:evolutionary divergence in the genus *Drosophila*. J Mol Evol 24:174–188.

Lucas P, Renou M (1989) Responses to pheromone compounds in *Mamestra suasa* (Lepidoptera:Noctuidae) olfactory neurones. J Insect Physiol 35:837–845.

Lukefahr MJ, Houghtaling JE, Cruhm DG (1975) Suppression of *Heliothis* spp. with cottons containing combinations of resistant characters. J Econ Entomol 68:743–746.

Luttrell RG, Roush RT, Ali A, Mink JS, Reid MR, Snodgrass GL (1987) Pyrethroid resistance in field populations of *Heliothis virescens* (Lepidoptera:Noctuidae) in Mississippi in 1986. J Econ Entomol 80:985–989.

Lysenko O (1963) The mechanisms of pathogenicity of *Pseudomonas aeruginosa* (Schroeter) Migula 1. The pathogenicity of strain N-06 for larvae of the greater wax moth, *Galleria mellonella* (Linnaeus). J Insect Pathol 5:78–82.

Mabbett TH, Dareepat P, Nachapong M (1980) Behaviour studies on *Heliothis armigera* and their application to scouting techniques for cotton in Thailand. Trop Pest Mgmt 26:268–273.

Mackay DA (1985) Conspecific host discrimination by ovipositing *Euphydryas editha* butterflies: its nature and its consequences for offspring survivorship. Res Pop Ecol 27:87–98.

Mackay DA, Jones RE (1989) Leaf shape and the host-finding behaviour of two ovipositing monophagous butterfly species. Ecol Ent 14:423–431.

Madsen BM and Miller LA (1987) Auditory input to motor neurons of the dorsal longitudinal muscles in a noctuid moth (*Barathra brassicae* L.) J Comp Physiol A 160:23–31.

Mahon RJ, Miethke PM, Mahon JA (1982) The evolutionary relationships of the Jarrah leaf miner, *Perthida glyphopa* (Common) (Lepidoptera:Incurvariidae). Aust J Zool 30:243–249.

Maniatis T, Fritsch EF, Sambrook J (1982) *Molecular cloning – A Laboratory Manual*. Cold Spring Harbor: Cold Spring Harbour Laboratory.

Marsh D, Kennedy JC, Ludlow AR (1978) An analysis of anemotactic zigzagging flight in male moths stimulated by pheromone. Physiol Ent 3:221–240.

Marsh PM (1978) The braconid parasites (Hymenoptera) of *Heliothis* species (Lepidoptera:Noctuidae). Proc Ent Soc Wash 80:15–36.

Marston NL, Dickerson WA, Ponder WW, Booth GD (1979) Calibration ratios for sampling soybean Lepidoptera: Species, larval size, plant growth stage and individual sampler. J Econ Entomol 72:110–114.

Martin PB, Lingren PD, Greene GL (1976) Relative abundance and host preferences of cabbage looper, soybean looper, tobacco budworm, and corn earworm on crops grown in Northern Florida. Env Ent 5:878–882.

Mather K, Jinks JL (1977) *Introduction to Biometrical Genetics*. Ithaca: Cornell University Press.

Mathews M (1987) The classification of *Heliothinae* (Noctuidae). PhD Thesis. British Museum and Kings College, London. 253 pp.

May RM (1973) *Stability and Complexity in Model Ecosystems*. Princeton: Princeton University Press.

May RM (1981) Models for interacting populations. In: RM May (Ed), *Theoretical Ecology: Principles and Applications*, Oxford: Blackwell Scientific Publications, pp. 5–29.

May RM (1986) The search for patterns in the balance of nature: advances and retreats. Ecology 67:1115–1126.

Mayr E (1942) *Systematics and the Origin of Species*. New York: Columbia University Press.

Mayr E (1963) *Animal Species and Evolution*. Cambridge, MA: Harvard University Press.

Mayr E (1969) The biological meaning of species. Biol J Linn Soc 1:311–320.

Mayse MA, Kogan M, Price PW (1978a) Sampling abundances of soybean arthropods: Comparison of methods. J Econ Entomol 71:135–141.

Mayse MA, Price PW, Kogan M (1978b) Sampling methods for arthropod colonization studies in soybean. Can Ent 110:265–274.

Maywald GF, O'Neill BM, Taylor MJF, Baillie SA (1985) A low-cost microprocessor based field data logger. CSIRO Div Ent Rep No. 36.

McCaffery AR, Maruf GM, Walker AJ, Styles K (1988) Resistance to pyrethroids in *Heliothis* spp.: Bioassay methods and incidence in populations from India and Asia. *Brighton Crop Protection Conference, 1988*. pp. 433–438.

McLaren IW (1981) The use of *Trichogramma* spp. in tomatoes. In: PH Twine (Ed), *Workshop on Biological Control of* Heliothis. Brisbane: Qld DPI.

McKenzie JA, Whitten MJ, Adena MA (1982) The effect of genetic background on the fitness of diazinon resistance genotypes of the Australian sheep blowfly, *Lucilia cuprina*. Heredity 49:1-9.

Meng LZ, Wu CH, Wicklein M, Kaissling KE, Bestmann HJ (1989) Number and sensitivity of three types of pheromone receptor cells in *Antheraea pernyi* and *A. polyphemus*. J Comp Physiol A 165:139-146.

Messenger PS, Flitters NE (1958) Effect of constant temperature environments on the egg stage of three species of Hawaiian fruit flies. Ann Ent Soc Am 51:109-119.

Meteorological Office (1968) *The Measurement of Upper Winds by Means of Pilot Balloons*. 4th ed. London: Her Majesty's Stationery Office.

Mikkola K (1971) Pollen analysis as a means of studying the migrations of Lepidoptera. Ann Ent Fenn 37:136-139.

Miller JR, Miller TA (Eds) (1986) *Insect-Plant Interactions*. Springer Series in Experimental Entomology, New York: Springer-Verlag.

Miller JR, Roelofs WL (1978) Gypsy moth responses to pheromone enantiomers as evaluated in a sustained flight wind tunnel. Env Ent 4:42-44

Miller JR, Strickler KL (1984) Finding and accepting host plants. In: WJ Bell, RT Carde (Eds), *The Chemical Ecology of Insects*, London: Chapman and Hall, pp. 211-228.

Miller LA (1983) How insects detect and avoid bats. In: F Huber, H Markl (Eds), *Neuroethology and Behavioural Physiology*, Berlin: Springer, pp.251-266.

Miller SG, Huettel MD, Davis MTB, Weber EH, Weber LA (1986) Male Sterility in *Heliothis virescens* x *H.subflexa* backcross hybrids. Evidence for abnormal mitochondrial transcripts in testis. Mol Gen Genet 203:451-461.

Mitchell ER, Webb JC, Hines RW, Stanley JW, Endris RG, Lindquist DA, Masuda, S (1972) Evaluation of cylindrical electric grids as pheromone traps for loopers and tobacco hornworms. Env Ent 1:365-368.

Moran C, Shaw DD (1977) Population cytogenetics of the genus *Caledia* (Orthoptera:Acridinae) III. Chromosomal polymorphism, racial parapatry an introgression. Chromosoma 63:181-204.

Morton R, Tuart LD, Wardhaugh KG (1981) The analysis and standardisation of light-trap catches of *Heliothis armiger* (Hübner) and *H. punctiger* Wallengren (Lepidoptera:Noctuidae) Bull Ent Res 71:207-225.

Mueller TF, Harris VE, Phillips JR (1984) Theory of *Heliothis* (Lepidoptera:Noctuidae) management through reduction of the first spring generation. Env Ent 13:625-634.

Muirhead-Thomson RC (1948) Studies on *Anopheles gambiae* and *A. melas* in and around Lagos. Bull Ent Res 38:527-558.

Muirhead-Thomson RC (1951) Studies on salt-water and fresh-water *Anopheles gambiae* on the East African coast. Bull Ent Res 41:487-501.

Murdoch WW, Chesson G, Chesson PL (1985) Biological control in theory and practice. Am Nat 125:344-366.

Murlis J (1986) The structure of odour plumes. In: TL Payne, MC Birch, CEJ Kennedy (Eds), *Mechanisms in Insect Olfaction*, Oxford: Clarendon Press, pp. 27-38.

Murlis J, Bettany BW (1977) Night flight towards a sex pheromone source by male *Spodoptera littoralis* (Boisd.) (Lepidoptera, Noctuidae). Nature 268:433–435.

Murlis J, Bettany BW, Kelley J, Martin L (1982) The analysis of flight paths of male Egyptian cotton leafworm moths, *Spodoptera littoralis*, to a sex pheromone source in the field. Physiol Entomol 7:435–441.

Murray D, Wicks R (1986) *Microplitis* spp.: A promising parasite. In: MP Zalucki, PH Twine (Eds), *Proceedings of the* Heliothis *Ecology Workshop 1985*. Brisbane: Qld DPI publication, pp. 151–152.

Mustaparta H (1984) Olfaction. In: WS Bell, RT Carde (Eds), *Chemical Ecology of Insects*, London: Chapman and Hall, pp.37–72.

Nadgauda D, Pitre H (1983) Development, fecundity, and longevity of the tobacco budworm (Lepidoptera:Noctuidae) fed soybean, cotton, and artificial diet at three temperatures. Env Ent 12:582–586.

Nair KSS, McEwen SL (1976) Host selection by the adult cabbage root fly, *Hylemya brassicae* (Diptera:Anthomyiidae); effect of glucosinolates and common nutrients on oviposition. Can Ent 108:1021–1030.

Nei M (1978) Estimation of average heterozygosity and genetic distance from a small number of individuals. Genetics 89:583–590.

Nemec SJ (1971) Effects of lunar phase on light-trap collections and populations of bollworm moths. J Econ Entomol 64:860–864.

Nye IWB (1982) The nomenclature of *Heliothis* and associated taxa (Lepidoptera: Noctuidae):Past and present. In: W Reed and V Kumble (Eds), *Proceedings of the International Workshop on Heliothis Management*, 15–20 November 1981, ICRISAT Center, Patancheru, A.P., India.

Oatman ER, Platner GR (1978) Effects of mass release of *Trichogramma pretiosum* against lepidopterous pests on processing tomatoes in Southern California, with notes on host egg population trends. J Econ Entomol 71:896–900.

O'Connell RJ, Grant AJ, Mayer MS, Mankin RW (1983) Morphological correlates of differences in pheromone sensitivity in insect sensilla. Science 220:1408–1410.

Odum EP (1971) *Fundamentals of Ecology*. Philadelphia: W.B. Saunders.

Oku T (1980) Evening activity of the pre-reproductive adults of two Euxoa cutworms (Lepidoptera:Noctuidae) Appl Ent Zool 15:344–347.

Oku T (1983) Annual and geographical distribution of crop infestation in northern Japan by the oriental armyworm in special relation to the migration phenomenon. (In Japanese, with English summary.) Misc Publ Tohoku Nat Agric Exp Station No. 3:1–49. (Rev Appl Ent (A) 71:854–855).

Oldberg RM (1983) Pheromone-triggered flip-flopping interneurons in the ventral nerve cord of the silkworm moth, *Bombyx mori*. J Comp Physiol A. 152:297–307.

Olsen P, Nielsen ES, Skule B (1984) En batteridrevet letvaegts lysrorsfaelde tilindsamling af natflyvende insekter. Lepidoptera 7:237–248.

Oyeyele S, Zalucki MP (1990) Cardiac glycosides and oviposition by *Danaus plexippus* on *Asclepias fruticosa* in south-east Queensland (Australia), with notes on the effect of plant nitrogen content. Ecol Ent 15:177–185.

Pal R, Whitten MJ (Eds) (1974) *The Use of Genetics in Insect Control*. Amsteram: Elsevier North-Holland.

Papaj DR, Rausher MD (1983) Individual variation in host location by phytophagous insects. In: S Ahmad (Ed), *Herbivorous Insects. Host-Seeking Behavior and Mechanisms*, New York: Academic Press, pp. 27–38.

Papaj D, Prokopy RJ (1989) Ecological and evolutionary aspects of learning in phytophagous insects. Ann Rev Entomol 34:315–350.

Pashley DP (1986) Host-associated genetic differentiation in fall armyworm (Lepidoptera:Noctuidae): A sibling species complex? Ann Ent Soc Am 79:898–904.

Pashley DP, Bush GL (1979) The use of allozymes in studing insect movement with special reference to the codling moth, *Laspeyresia pomonella* (L.) (Olethreutidae) In: RL Rabb, GG Kennedy (Eds), *Movement of Highly Mobile Insects: Concept and Methodology in Research*. North Carolina: University of Graphics, Raleigh. pp. 333–341.

Pashley DP, Johnson SJ, Sparks AN, Jr. (1985) Genetic population structure of migratory moths: The fall armyworm (Lepidoptera:Noctuidae) Ann Ent Soc Am 78:756–762.

Pashley DP, Proverbs MD (1981) Quality control by electrophoretic monitoring in a laboratory colony of codling moths. Ann Ent Soc Am 74:20–23.

Paterson HEH (1956) Status of the two forms of housefly occurring in South Africa. Nature 178:928–929.

Paterson HEH (1962) Status of the East African salt-water-breeding variant of *Anopheles gambiae* Giles. Nature 195:469–470.

Paterson HEH (1978) More evidence against speciation by reinforcement. SA J Sci 74:369–371.

Paterson HEH (1982) Perspective on speciation by reinforcement. SA J Sci 78:53–57.

Paterson HEH (1985) The recognition concept of species. In: E Vrba (Ed), *Species and Speciation*. Transvaal Museum, Pretoria: Transvaal Museum Monograph No. 4.

Paul DH (1974) Responses to acoustic stimualtion of thoracic interneurons in noctuid moths. J Insect Physiol 20:2205–2218.

Payne CD (1985) *The GLIM System Release 3.77 Manual*. Oxford: Numerical Algorithims Group.

Payne RW (1987) *Genstat 5 Reference Manual*. Lawes Agricultural Trust. Oxford: Clarendon Press.

Payne TL, Birch MC, Kennedy CEJ (Eds) (1986) *Mechanisms in Insect Olfaction*, Oxford: Clarendon Press.

Pearson EO (1958) *The Insect Pests of Cotton in Tropical Africa*. London: Commonwealth Institute of Entomology.

Pedgley DE (1980) Weather and airborne organisms. World Meteorological Organisation Technical Note No. 173: Geneva: World Meteorological Organisation.

Pedgley DE (1985) Windborne migration of *Heliothis armigera* (Hübner) (Lepidoptera: Noctuidae) to the British Isles. Entomologist's Gazette 36:15–20.

Pedgley DE (1986) Windborne migration in the Middle East by the moth *Heliothis armigera* (Lepidoptera:Noctuidae) Ecol Ent 11:467–470.

Pedgley DE, Reynolds DR, Riley JR Tucker MR (1982) Flying insects reveal smallscale wind systems. Weather 37:295–306.

Pedigo LP, Lentz GL, Stone JD, Cox DF (1972) Green cloverworm populations in Iowa soybean with special reference to sampling procedure. J Econ Entomol 65:414–421.

Perry JN, Wall C (1986) The effect of habitat on the flight of moths orienting to pheromone sources. In: TL Payne, MC Birch, CEJ Kennedy (Eds), *Mechanisms in Insect Olfaction*, Oxford: Clarendon Press, pp. 91–96.

Persson B (1976) Influence of weather and nocturnal illumination on the activity and abundance of populations of Noctuids (Lepidoptera) in south coastal Queensland. Bull Ent Res 66:33–63.

Phillips JR (1979) Migration of the bollworm, *Heliothis zea* (Boddie) In: RL Rabb, GG Kennedy (Eds), *Movement of Highly Mobile Insects: Concepts and Methodology in Research*. Raleigh, North Carolina: University of Graphics, pp.409–411.

Phillips JR, Newsom LD (1966) Diapause in *Heliothis zea* and *Heliothis virescens* (Lepidoptera: Noctuidae) Ann Ent Soc Am 59:154–159.

Pickens LG, Thimijan RW (1986) Design parameters that affect the performance of UV-emitting traps in attracting house flies (Diptera:Muscidae). J Econ Entomol 79:1003–1009.

Plapp FW Jr (1979) Synergism of pyrethroid insecticides by formaminides against *Heliothis* pests of cotton. J Econ Entomol 72:667–670.

Plapp FW, McWhorter GM, Vance WH (1987) Monitoring for pyrethroid resistance in the tobacco budworm in Texas – 1986. In: *Proceedings of Beltwide Cotton Production Research Conferences*, Dallas, Texas, pp. 324–326.

Podoler H, Rogers D (1975) A new method for the identification of key factors from life table data. J Anim Ecol 44:84–114.

Podolsky AS (1984) *New Phenology. Elements in Mathematical Forecasting*. New York: John Wiley & Sons, Inc.

Poinar GO (1975) Description and biology of a new insect parasitic rhabditoid, *Heterorhabditis bacteriophora* n. gen., n. sp. (Rhabditida, Heterorhabditidae n. fam.) Nematologia 21:463–470.

Powell JE, King EG (1984) Behaviour of adult *Microplitis croceipes* and parasitism of *Heliothis* spp host larvae in cotton. Env Ent 13:272–277.

Pradhan S (1945) Insect population studies. II. Rate of insect development under variable temperatures of the field. Proc Nat Inst Sci India 11:74–80.

Pradhan S (1946) Insect population studies. IV. Dynamics of temperature effect on insect development. Proc Nat Inst Sci India 12:385–404.

Pretorius LM (1976) Laboratory studies on the development and reproductive performance of *Heliothis armigera* (Hubn.) on various food plants. J Ent Soc Sth Afr 39:337–343.

Priesner E (1979) Specificity studies on pheromone receptors of noctuid and tortricid Lepidoptera. In: FJ Ritter (Ed), *Chemical Ecology: Odour Communication in Animals*, Amsterdam: Elsevier, pp. 57–71.

Prokopy RJ, Owen ED (1983) Visual detection of plants by herbivorous insects. Ann Rev Entomol 28:337–364.

Prokopy RJ, Averill AL, Cooley SS, Roitberg CA (1982) Associative learning in egglaying site selection by apple maggot flies. Science 218:76–77.

Proshold FI (1983) Release of backcross insects on St. Croix, U.S. Virgin Islands, to suppress the tobacco budworm (Lepidoptera:Noctuidae): Infusion of sterility into a native population. J Econ Entomol 76:1353–1359.

Proshold FI, Laster ML, Martin DF, King EG (1982) The potential for hybrid sterility in *Heliothis* management. In: W Reed, and V Kumble (Eds) *Proceed-*

ings of the International Workshop on Heliothis Management. 15–20 November 1981, ICRISAT Center, Patancheru, A.P., India. p.170.

Proshold FI, Raulston JR, Martin DF, Laster ML (1983) Release of backcross insect on St. Croix to suppress the tobacco budworm (Lepidoptera:Noctuidae): Behavior and interaction with native insects. J Econ Entomol 76:626–631.

Pruess KP, Pruess NC (1971) Telescopic observation of the moon as a means for observing migration of the army cutworm, *Chorizagrotis auxiliaris* (Lepidoptera:Noctuidae) Ecology 52:999–1007.

Puterka GJ, Slosser JE, Price JR (1985) Parasites of *Heliothis* spp: Parasitism and seasonal occurrence for host crops in the Texas rolling plains. Env Ent 14:441–446.

Pyke B (1981) Some observations on naturally-occurring predators in insecticidaly treated cotton fields in South-east Queensland. 1975–1979. In: PH Twine (Ed), *Workshop on Biological Control of* Heliothis. Brisbane: QLD DPI publication.

Qayyum A, Zalucki MP (1987) Effects of high temperature on survival of eggs of *Heliothis armigera* (Hübner) and *H. punctigera* Wallengren (Lepidoptera:Noctuidae) J Aust Ent Soc 26:295–296.

Quinn TW, White BN (1987) Identification of restriction-fragment-length- polymorphisms in genomic DNA of the lesser snow goose (*Anser caerulescens caerulescens*) Mol Biol Evol 4:126–143.

Rabb RL (1979) Regional research on insect movement: initial considerations. In: RL Rabb, GG Kennedy (Eds), *Movement of Highly Mobile Insects: Concepts and Methodology in Research.* Raleigh, North Carolina: University Graphics, pp. 2–12.

Rainey RC (1976) Flight behaviour and features of the atmospheric environment. Symp Roy Entomol Soc London 7:75–112.

Ramaswamy SB (1988) Host finding by moths: Sensory modalities and behaviours. J Insect Physiol 34:235–249.

Ramaswamy SB, Ma WK, Baker GT (1987) Sensory cues and receptors for oviposition by *Heliothis virescens.* Ent Exp et Appl 43:159–68.

Ramaswany SB, Roush RT, Kitten WF (1985) Release and recapture probabilities of laboratory-adapted and wild-type *Heliothis virescens* (F.) (Lepidoptera: Noctuidae) in pheromone baited traps. J Ent Sci 20:460–464.

Ratte HT (1985) Temperature and insect development. In: KH Hoffman (Ed), *Environmental Physiology and Biochemistry of Insects.* Berlin: Springer-Verlag, pp. 33–66.

Raulston JR (1979) Tagging of natural populations of Lepidoptera for studies of dispersal and migration. In: RL Rabb, GG Kennedy (Eds), *Movement of Highly Mobile Insects: Concepts and Methodology in Research.* North Carolina: University Graphics, Raleigh, pp.354–358.

Raulston JR, Sparks AN, Lingren PD (1980) Design and comparative efficiency of a wind-orientated trap for capturing live *Heliothis* spp. J Econ Entomol 73:586–589.

Raulston JR, Wolf WW, Lingren PD, Sparks AN (1982) Migration as a factor in *Heliothis* management. In: W Reed, V Kumble (Eds), *Proceedings of the International Workshop on Heliothis Management.* 15–20 November 1981, ICRISAT Center, Patancheru, A.P., India.

Rausher MD (1980) Host abundance, juvenile survival, and oviposition preference in *Battus philenor*. Evolution 34:342–355.

Reed W (1965) *Heliothis armigera* (Hb.) (Noctuidae) in western Tanganyika I – Biology, with special reference to the pupal stage. Bull Ent Res 56:117–126.

Reling D, Taylor RAJ (1984) A collapsible tow net used for sampling arthropods by airplane. J Econ Entomol 77:1615–1617.

Rembold H, Tober H (1987) Kairomones in legumes and their effect on *Heliothis armigera*. In: V Labeyrie, G Fabres, D Lachaise (Eds), *Insect – Plant. Proc. 6th Int. Symp. on Insect-Plant Relationships* (Pau 1986) Dr. W Junk, Dordrecht, pp. 25–30.

Renwick JA, Radke CD (1985) Constituents of host and non-host plants deterring oviposition by the cabbage butterfly, *Pieris rapae*. Ent Exp et Appl 39:21–26.

Richardson BJ, Baverstock PR, Adams M (1986) *Allozyme Electrophoresis. A Handbook for Animal Systematics and Population Studies*. Sydney: Academic Press Australia.

Riley JR (1989) Remote sensing in entomology. Ann Rev Entomol 34:247–271.

Riley JR, Reynolds DR (1979) Radar-based studies of the migratory flight of grasshoppers in the middle Niger area of Mali. Proc R Soc Lond B 204:67–82.

Riley JR, Reynolds DR, Farmery MJ (1981) Radar observations of *Spodoptera exempta* Kenya, March–April 1979. Centre for Overseas Pest Research, London. Misc Rep No. 54.

Riley JR, Reynolds DR, Farmery MJ (1983) Observations of the flight behaviour of the armyworm moth, *Spodoptera exempta*, at an emergence site using radar and infra-red optical techniques. Ecol Ent 8:395–418.

Riley JR, Reynolds DR, Farrow RA (1987) Migration of *Nilaparvata lugens* (Stal) (Delphacidae) and other Hemiptera associated with rice during the dry season in the Philippines:a study using radar, visual observations, aerial netting and ground trapping. Bull Ent Res 77:145–169.

Riley JR, Smith AD, Bettany DW (1990) The use of video equipment to record in three dimensions the flight trajectories of *Heliothis armigera* and other moths at night. Physiol Entomol 15:73–80.

Rind CF (1983) The organisation of flight motoneurones in the moth, *Manduca sexta*. J Exp Biol 102:239–251.

Riordan AJ (1979) Density and availability of meteorological data and their use in illustrating atmospheric long-range transport in the southeastern United States. In: RL Rabb, GG Kennedy (Eds), *Movement of Highly Mobile Insects: Concepts and Methodology in Research*. Raleigh, North Carolina: University Graphics, pp.120–132.

Roach SH (1981) Emergence of overwintered *Heliothis* spp. moths from three different tillage systems. Env Ent 10:817–818.

Robinson HS (1952) On the behaviour of night flying insects in the neighbourhood of a bright light source. Proc Roy Ent Soc Lond (A) 27:13–21.

Robinson HS, Robinson PJM (1950) Some notes on the observed behaviour of Lepidoptera in flight in the vicinity of light-sources together with a description of a light-trap designed to take Entomological samples. Ent Gaz 1:3–15.

Roeder KD (1951) Movements of the thorax and potential changes in the thoracic muscles of insects during flight. Biol Bull 100:95–106.

Roeder KD (1965) Moths and ultrasound. Sci Am 212: 94–102.

Roeder KD (1966) Interneurons of the thoracic nerve cord activated by tympanic nerve fibres in noctuid moths. J Insect Physiol 12: 1227–1244.

Roeder KD (1967) Turning tendency of moths exposed to ultrasound while in stationary flight. J Insect Physiol 13:873–888.

Roeder KD, Payne RS (1966) Acoustic orientation of a moth in flight by means of two sense cells. Symp Soc Exp Biol 20:251–272.

Roeder KD, Treat AE (1957) Ultrasonic reception by the tympanic organ of noctuid moths. J Exp Zool 134:127–157.

Roelofs WL (1977) An overview – the evolving philosophies and methodologies. In: HH Shorey, JJ McKelvey (Eds), *Chemical Control of Insect Behavior. Theory and Application*, New York: Wiley-Interscience, pp. 287–298.

Rogers DJ (1982) Screening legumes for resistance to *Heliothis*. In: W Reed, V Kumble (Eds), *Proceedings of the International Workshop on Heliothis Management*, November 15–20, 1981. ICRISAT Centre, Patancheru, A.P., India. pp 277–287.

Rogers JS (1972) Measures of genetic similarity and genetic distance. In: Studies in Genetics. Univ Texas Publ 7213:145–153.

Roitberg BD, Prokopy RJ (1983) Host deprivation influence on response of *Rhagoletis pomonella* to its oviposition deterring pheromone. Physiol Ent 8:69–72.

Room PM (1977) 32P-labelling of immature stages of *Heliothis armigera* (Hübner) and *H. punctigera* Wallengren (Lepidoptera:Noctuidae): relationships of dose to radioactivity, mortality and label half-life. J Aust Ent Soc 16:245–251.

Room PM (1978) A prototype 'on-line' system for management of cotton pests in the Namoi Valley, New South Wales. Prot Ecol 1:245–264.

Room PM (1979) Parasites and predators of *Heliothis* spp. (Lepidoptera:Noctuidae) in cotton in the Namoi Valley, New South Wales. J Aust Ent Soc 18:223–228.

Room PM (1983) Calculations of temperature-driven development by *Heliothis* spp. (Lepidoptera:Noctuidae) in the Namoi Valley, New South Wales. J Aust Ent Soc 22:211–215.

Roome RE (1979) Pupal diapause in *Heliothis armigera* (Hübner) (Lepidoptera:Noctuidae) in Botswana: its regulation by environmental factors. Bull Ent Res 69:149–160.

Rose DJW, Page WW, Dewhurst CF, Riley JR, Reynolds DR, Pedgley DE, Tucker, MR (1985) Downwind migration of the African armyworm moth, *Spodoptera exempta*, studied by mark-and-capture and by radar. Ecol Ent 10:299–313.

Rothschild GHL (1966) A study of a natural population of *Conomelus anceps* Germar (Homoptera:Delphacidae) including observations on predation using the precipition test. J Anim Ecol 35:413–433.

Rothschild GHS (1978) Attractants for *Heliothis armigera* and *H. punctigera*. J Aust Ent Soc 102:389–390.

Rothschild GHS, Wilson AG, Malafant K (1982) Preliminary studies on the female sex pheromones of *Heliothis* species and their possible use in control programs in Australia. In: W Reed, V Kumble (Eds), *Proceedings of the International Workshop on Heliothis Management*, ICRISAT Center, Patancheru, A.P., India 15–20 November, 1981, pp. 319–328.

Roughgarden J (1979) *Theory of Population Genetics and Evolutionary Ecology: An Introduction*. New York: Macmillan.

Roush RT, Daly JC (1990) The role of population genetics in resistance research and management. In: RT Roush, B Tabashnik (Eds), *Pesticide Resistance in Arthropods*. New York: Chapman and Hall, pp. 97–152.

Roush RT, Hoy MA (1981) Laboratory, glasshouse and field studies of artificially selected carbaryl resistance in *Metaseiulus occidentalis*. J Econ Entomol 74: 142–147.

Roush RT, McKenzie JA (1987) Ecological genetics of insecticide and acaricide resistance. Ann Rev Entomol 32:361–380.

Roush RT, Miller GL (1986) Considerations for the design of insecticide resistance monitoring programs. J Econ Entomol 79:293–298.

Royama T (1980) Effect of adult dispersal on the dynamics of local populations of an insect species: A theoretical investigation. In: AA Berryman, L Safranyid (Eds), *Dispersal of Forest Insects: Evaluation, Theory and Management*. Washington: Cooperative Extension Service, State University, Pullman, Washington, pp. 79–93.

Royama T (1981) Evaluation of mortality factors in insect life table analysis. Ecol Mono 51:495–505.

Rumbo ER (1981) Study of single sensillum responses to pheromone in the light-brown apple moth, *Epiphyas postvittana*, using an averaging technique. Physiol Entomol 6:87–98.

Sage TL, Gregg P (1985) A comparison of four types of pheromone traps for *Heliothis armigera* (Hübner) (Lepidoptera:Noctuidae). J Aust Ent Soc 24:99–100.

Samson PR, Blood PR (1980) Voracity and searching ability of *Chrysopa signata* (Neuroptera:Chrysopidae), *Micromus tasmaniae* (Neuroptera:Hemerobiidae) and *Tropiconabis capsiformis* (Hemiptera:Nabidae). Aust J Zool 28:575–580.

Saxena KN, Rembold H (1984) Attraction of *Heliothis armigera* (Hübner) larvae by chickpea seed powder constituents. Zeitschrift fur angewandte Entomologie 97:145–153.

Saxena KN, Rembold H (1985) Orientation and ovipositional responses of *Heliothis armigera* to certain neem constituents. In: H Schmutterer, KRS Ascher (Eds), *Proceedings of the 2nd International Neem Conference*, Rauischholzhausen, 1983. GTZ, Eschborn.

Saxena KN, Khattar P, Goyal S (1976) Measurement of orientation responses of caterpillars indoors and outdoors on a grid. Experientia 33:1312–1313.

Sayer HJ (1956) A photographic method for the study of insect migration. Nature 177:226.

Schaefer GW (1976) Radar observations of insect flight. Proc Symp R Ent Soc Lond 7:157–197.

Schaefer GW (1979) An airborne radar technique for the investigation and control of migrating insect pests. Phil Trans R Soc Lond B 287:459–465.

Schaefer GW, Bent GA (1984) An infra-red remote sensing system for the active detection and automatic determination of insect flight trajectories (IRADIT). Bull Ent Res 74:261–278.

Schneider D (1957) Elektrophysiologische Untersuchungen von Chemo- und Mechanorezeptoren der Antenne des Seidenspinners *Bombyx mori* L. Zeitschrift Vergl Physiol 40:8–41.

Schneider JC, Benedict JH, Gould F, Meredith WR, Schuster MF, Zummo GR (1986) Interaction of *Heliothis* with its host plants. In: SJ Johnson, EG King, JR Bradley (Eds), *Theory and Tactics of Heliothis Population Management*: 1 — *Cultural and Biological Control*. Southern Cooperative Series Bulletin 316, Stillwater: Oklahoma State University, pp. 2–21.

Schoonhoven LM (1987) What makes a caterpillar eat? The sensory code underlying feeding behaviour. In: RF Chapman, EA Bernays, JG Stoffolano (Eds), *Perspectives in Chemoreception and Behaviour*, London: Springer-Verlag, pp. 69–97.

Schuster MF, Anderson RE (1976) Insecticidal efficacy of insect resistant cotton. J Econ Entomol 69:691–692.

Scott RW, Achtemeier GL (1987) Estimating pathways of migrating insects. Env Ent 16:1244–1254.

Seber GAF (1973) *The Estimation of Animal Abundance and Related Parameters*. London: Griffin.

Sell DK, Whitt GS, Metcalf RL, Lee L-p K. (1974) Enzyme polymorphism in the corn earworm, *Heliothis zea* (Lepidoptera:Noctuidae) Hemolymph esterase polymorphism. Can Ent 106:701–709.

Sell DK, Whitt GS, Luckmann WH (1975) Esterase polymorphism in the corn earworm, *Heliothis zea* (Boddie):a survey of temporal and spatial allelic variation in natural populations. Biochem Genet 13:885–898.

Seymour JE (1985) The ecology of *Microgaster demolitor* (Hymenoptera:Braconidae), a larval parasitoid of *Heliothis* spp. (Lepidoptera:Noctuidae) Honours thesis, James Cook University, Townsville, Australia.

Sharpe PJH and DeMichele DW (1977) Reaction kinetics of poikilotherm development. J Theor Biol 64:649–670.

Sharpe PJH, Schoolfield RM and Butler GD (1981) Distribution model of *Heliothis zea* (Lepidoptera:Noctuidae) development times. Can Ent 113:845–856.

Shaw DD (1976) Population cytogenetics of the genus *Caledia* (Orthoptera:Acridinae) I. Inter-and Intraspecific karyotype diversity. Chromosoma 54:221–243.

Sherlock PL, Bowden J, Digby PGN (1985) Studies of elemental composition as a biological marker in insects. IV. The influence of soil type and host-plant on elemental composition of *Agrotis segetum* (Denis & Schiffermuller) (Lepidoptera:Noctuidae). Bull Ent Res 75:675–687.

Shiller I (1946) A hibernation cage for the pink bollworm. US Bur Pl Quarantine, E T Ser No. 226, 1–6.

Shorey HH, Hale RL (1965) Mass-rearing of the larvae of nine noctuid species on a simple artificial medium. J Econ Entomol 58:522–524.

Shumakov EM, Yakhimovich LA (1955) Morphological and histological perculiarities in the metamorphosis of the cotton bollworm *Chloridea obsoleta* F. in connection with the phenomenon of diapause. Rev Appl Ent 44:11–12.

Singer MC (1982) Quantification of host preferences by manipulation of oviposition behaviour in the butterfly *Euphydryas editha*. Oecologia 52:224–229.

Singer MC (1983) Determinants of multiple host use by a phytophagous insect population. Evolution 37:389–403.

Singer MC (1986) The definition and measurement of oviposition preference. In: JR Miller, TA Miller (Eds), *Insect-Plant Interactions*. Springer Series in Experimental Entomology, New York: Springer-Verlag, pp. 65–94.

Singer MC, Wedlake P (1981) Capture does affect probability of recapture in a butterfly species. Ecol Ent 6:215–216.

Slatkin M (1981) Estimating levels of gene flow in natural populations. Genetics 99:323–335.

Slatkin M (1985) Gene flow in natural populations. Ann Rev Ecol Syst 16:393–430.

Slatkin M (1987) Gene flow and the geographic structure of natural populations. Science 236:787–792.

Sloan WJS (1938) The maize trap crop for the control of corn earworm in cotton. Qld Agric J January 1938, p. 76.

Slosser JE, Phillips JR, Herzog GA, Reynolds CR (1975) Overwinter survival and spring emergence of the bollworm in Arkansas. Env Ent 4:1015–1024.

Sluss TP, Graham HM (1979) Allozyme variation in natural populatons of *Heliothis virescens*. Ann Ent Soc Am 72:317–322.

Sluss TP, Sluss ES, Graham HM, Dubois M (1978) Allozyme differences between *Heliothis virescens* and *H. zea*. Ann Ent Soc Am 71:191–195.

Smith AM, McDonald G (1986) Interpreting ultraviolet-light and fermentation trap catches of *Mythimna convecta* (Walker) (Lepidoptera:Noctuidae) using phenological simulation. Bull Ent Res 76:419–431.

Smith CAB (1970) A note on testing the Hardy–Weinberg law. Ann Hum Genet 33:377–383.

Smith LS, Smith DA (1982) The Naturalized *Lantana camara* complex in eastern Australia. Qld Bot Bull 1:1–26.

Smith PW, Taylor JG, Apple JW (1959) A comparison of insect traps equipped with 6- and 15-watt blacklight lamps. J Econ Entomol 52:1212–1214.

Sneath PHA, Sokal RR (1973) *Numerical Taxonomy*. San Francisco: Freeman.

Snow JW, Cantelo WW, Bowman MC (1969) Distribution of the corn earworm on St. Croix, U.S. Virgin Islands, and its relation to suppression programs. J Econ Entomol 62:606–611.

Sokal RR (1973) The species problem reconsidered. Syst Zool 22:360–374.

Solbreck C (1985) Insect migration strategies and population dynamics. Contrib Marine Sci 27(supp):641–662.

Solignac M, Monnerot M (1986) Race formation, speciation and introgression within *Drosophila simulans*, *D. mauritiana* and *D. sechellia* inferred from mitochondrial DNA analysis. Evolution 40:531–539.

Southern Co-operative Series Bulletin 280 (1983) *Host Plant Resistance Research. Methods for Insects, Diseases, Nematodes and Spider Mites in Cotton*. Chapters 9 and 10. Starkville: Mississippi State University.

Southwood TRE (1962) Migration of terretrial arthropods in relation to habitat. Biol Rev 37:171–214.

Southwood TRE (1978) *Ecological Methods*. London: Methuen.

Sparks AN, Wright RL, Hollingsworth JP (1967) Evaluation of designs and installations of electric insect traps to collect bollworm moths in Reeves county, Texas. J Econ Entomol 68:431–432.

Sparks AN, Jackson RD, Allen CL (1975) Corn earworms: Capture of adults in light traps on unmanned oil platforms in the Gulf of Mexico. J Econ Entomol 68:431–432.

Sparks AN, Raulston JR, Carpenter JE, Lingren PD (1982) The present status and potential for novel uses of pheromone to control *Heliothis*. In: W Reed, V

Kumble (Eds), *Proceedings of the International Workshop on Heliothis Management*, 15-20 November 1981, ICRISAT Center, Patancheru, A. P., India.

Sparks AN, Raulston JR, Lingren PD, Carpenter JE, Klun JA, Mullinx BG (1979) Field response of male *Heliothis virescens* to pheromonal stimuli and traps. Bull Ent Soc Am 25:268-274.

Speith PT (1974) Gene flow and genetic differentiation. Genetics 78:961-965.

Sprenkel RK, Brooks WM (1975) Artificial dissemination and epizootic initiation of *Nomuraea rileyi*, an entomogenous fungus of lepidopterous pests of soybean. J Econ Entomol 68:847-851.

Stadelbacher EA (1981) Role of early-season wild and naturalised host plants in the build up of the F_1 generation of *Heliothis zea* and *Heliothis virescens* in the delta of Mississippi. Env Ent 10:766-770.

Stadelbacher EA, Powell JE, King EG (1984) Parasitism of *Heliothis zea* and *H. virescens* larvae in wild and cultivated host plants in the delta of the Mississippi. Env Ent 13:1167-1172.

Stadler E (1982) Sensory physiology of insect-plant relationships — round-table discussion. In: JH Visser, AK Minks (Eds), *Proc. 5th International Symposium on Insect-Plant Relationships*, Pudoc 1982, Wageningen, pp. 81-91.

Stadler E (1984) Contact chemoreception. In: WS Bell, RT Carde (Eds), *Chemical Ecology of Insects*, London: Chapman and Hall, pp. 3-36.

Stamp NE (1982) Behaviour of parasitized aposematic caterpillars: Advantageous to the parasitoid or host? Am Nat 118:715-725.

Stanton ML (1982) Searching in a patchy environment:foodplant selection by *Colias Periphyle* butterflies. Ecology 63:839-853.

Steinbrecht RA (1980) Cryofixation without cryoprotectants. Freeze substitution and freeze etching of an insect olfactory receptor. Tissue Cell 12:73-100.

Steinhorst RK, Hunt HW, Innis GS, Haydock KP (1978) Sensitivity analysis of the ELM model. In: GS Innis (Ed), *Grassland Simulation*. Berlin: Springer-Verlag, pp.231-256.

Steinhouse EA (1959) *Serratia marcescens* Bizio as an insect pathogen. Hilgardia 28:351-380.

Sterling WL (1976) Sequential decision plans for the management of cotton arthropods in southeast Queensland. Aust J Ecol 1:265-274.

Stewart WW (1978) Functional connections between cells as revealed by dye-coupling with a highly fluorescent naphthalimide tracer. Cell 14:741-759.

Stinner RE, Gutierrez AP, Butler GD (1974a) An algorithm for temperature-dependent growth rate simulation. Can Ent 106:519-524.

Stinner RE, Rabb RL, Bradley JR (1974b) Population dynamics of *Heliothis zea* (Boddie) and *H. virescens* (F.) in North Carolina: A simulation model. Env Ent 31:163-168.

Stinner RE, Rabb RL, Bradley JR Jr (1977) Natural factors operating in the population dynamics of *Heliothis zea* in North Carolina. Proc XV Int Congr Entomol 622-642.

Stinner RE, Saks M, Dohse L (1986) Modelling of agricultural pest displacement. In: W Danthanarayana (Ed), *Insect Flight: Dispersal and Migration*, Berlin: Springer-Verlag, pp. 235-241.

Stinner RE, Bradley JR Jr, Roach SH, Hartstack AW, Lincoln CG (1979) Sampling

Heliothis species on native hosts and field crops other than cotton. In: *Economic Thresholds and Sampling of Heliothis species on Cotton, Corn, Soybeans and Other Host Plants*. Southern Co-operatives Series Bulletin No. 231.

Stone ND, Coulson N, Frisbie, RE, Loh DK (1986) Expert systems in Entomology: Three approaches to problem solving. Bull Ent Soc Am 32:161–166.

Sturgeon KB, Mitton JB (1986) Allozyme and morphological differentiation of mountain pine beetles *Dendroctonus pondersae* Hopkins (Coleoptera:Scolytidae) associated with host tree. Evolution 40:290–302.

Surlykke A, Miller LA (1982) Central branching of three sensory axons from a moth ear (*Agrotis segetum*, Noctuidae) J Insect Physiol 28:357–364.

Surlykke A, Larsen ON, Michelsen A (1988) Temporal coding in the auditory receptor of the moth ear. J Comp Physiol A 162:367–374.

Swofford DL, Selander RB (1981) BIOSYS-1: A FORTRAN program for the comprehensive analysis of electrophoretic data in population genetics and systematics. J Heredity 72:281–283.

Tanigoshi LK, Browne RW (1978) Influence of temperature on life table parameters of *Metaseiulus occidentalis* and *Tetranychus mcdanieli* (Acarina:Phytoseiidae, Tetranychidae). Ann Ent Soc Am 71:313–316.

Tanigoshi LK, Browne RW, Hoyt SC, Lagier RF (1976) Empirical analysis of variable temperature regimes on life stage development and population growth of *Tetranychus mcdanieli* (Acarina:Tetranychidae). Ann Ent Soc Am 69:712–716.

Taylor J, Padgham DE, Perfect TJ (1982) A Light-trap with upwardly directed illumination and temporal segregation of the catch. Bull Ent Res 72:669–673.

Taylor LR (1974) Insect migration, flight periodicity and the boundary layer. J Anim Ecol 43:225–238.

Taylor LR, Brown ES (1972) Effects of light-trap design and illumination on samples of moths in the Kenya highlands. Bull Ent Res 62:91–112.

Taylor LR, Taylor RAJ (1977) Aggregation, migration an population dynamics. Nature 265:415–421.

Taylor MFJ (1984) The dependence of development and fecundity of *Samea multiplicalis* on early larval nitrogen intake. J Insect Physiol 30:779–785.

Teakle RE (1973) A nuclear polyhedrosis virus from *Heliothis punctigera* Wallengren (Lepidoptera:Noctuidae) Qd J Agric Anim Sci 30:161–177.

Teakle RE (1974) A granulosis virus from *Heliothis punctigera*. J Invertebr Pathol 23:127–129.

Teakle RE (1977) Diseases of the *Heliothis* caterpillar. Qld Agric J 103:389–390.

Teakle RE, Jensen JM (1985) *Heliothis punctigera*. In: P Singh, RF Moore (Eds), *Handbook of Insect Rearing*. Vol 2. Amsterdam: Elsevier, pp.313–322.

Teal PE, McLaughlin JR, Tumlinson JH (1981a) Analysis of reproductive behaviour of *Heliothis virescens* under laboratory conditions. Ann Ent Soc Am 74:324–330.

Teal PE, Heath RR, Tumlinson JH, McLaughlin JR (1981b) Identification of a sex pheromone of *Heliothis subflexa* Gn. (Lepidoptera:Noctuidae) and field trapping studies using different blends of components. J Chem Ecol 7:1011–1022.

Teal PE, Tumlinson JH, Heath RR (1986) Chemical and behavioural analyses of volatile sex pheromone components released by calling *Heliothis virescen* (F.) females (Lepidoptera:Noctuidae). J Chem Ecol 12:107–126.

Tedders WL, Gottwald TR (1986) Evaluation of an insect collecting system and an ultra-low-volume spray system on a remotely piloted vehicle. J Econ Entomol 79:709–713.

Thorsteinson AJ (1960) Host selection in phytophagous insects. Ann Rev Entomol 5:193–218.

Tinbergen N, Meeuse BJD, Boerema LK, Varossieau WW (1942) Die Balz des Samfalters, *Eumenis* (*Satyrus*) *semele* (L.) Zeitschrift fur Tierpsychologie 5:182–226.

Tingle FC, Mitchell ER (1984) Aqueous extracts from indigenous plants as oviposition deterrents for *Heliothis* (F.) J Chem Ecol 10:101–113.

Titmarsh IJ (1985) Population dynamics of *Heliothis* spp. on tobacco in Far North Queensland. MSc Thesis, James Cook University, Townsville, Australia.

Tokeshi M (1985) Life-cycle and production of the burrowing mayfly, *Ephemera danica*: A new method for estimating degree-days required for growth. J Anim Ecol 54:919–930.

Topper CP (1978) The incidence of *Heliothis armigera* larvae and adults on ground nuts and sorghum and the prediction of oviposition on cotton. In *Proc 3rd Seminar for Cotton Pest Control in the Sudan*. Ciba Geigy, March 1978, Switzerland.

Topper CP (1987a) The dynamics of the adult population of *Heliothis armigera* (Hübner) (Lepidoptera:Noctuidae) within the Sudan Gezira in relation to cropping pattern and pest control on cotton. Bull Ent Res 77:525–539.

Topper CP (1987b) Nocturnal behaviour of adults of the *Heliothis armigera* (Hübner) (Lepidoptera:Noctuidae) in the Sudan Gezira and pest control implications. Bull Ent Res 77:541–554.

Traynier RMM (1979) Long-term changes in the oviposition behaviour of the cabbage butterfly, *Pieris rapae*, induced by contact with plants. Physiol Ent 4:87–96.

Treat AE (1979) Moth-borne mites and their hosts. In: RL Rabb, GG Kennedy (Eds), *Movement of Highly Mobile Insects: Concepts and Methodology in Research*. Raleigh, North Carolina: University Graphics, pp.359–368.

Tsukamoto M (1963) The log dosage-probit mortality curve in genetic research of insect resistance to insecticides. Botyu-Kagaku 29:91.

Tsukamoto M (1983) Methods of genetic analysis of insecticide resistance. In: GP Georghiou, T Saitos (Eds), *Pest Resistance to Pesticides*. New York: Plenum, pp.71–98.

Twine PH (1973) Egg parasites of *Heliothis punctigera* and *Heliothis armigera* in south-eastern Queensland. Qld J Agric Anim Sci 28:331–336.

Twine PH (1978) Effect of temperature on the development of larvae and pupae of the corn earworm, *Heliothis armigera* (Hübner) (Lepidoptera:Noctuidae) Qld J Agric Anim Sci 35:23–28.

Tyndale-Biscoe M (1984) Age-grading methods in adult insects:a review. Bull Ent Res 74:341–377.

Ullyett GC (1944) Unpublished Annual Report of the Parasite Laboratory, Pretoria: Plant Protection Research Institute.

Underwood T (1986) The analysis of competition by field experiments. In: J Kikkawa, DJ Anderson (Eds), *Community Ecology: Patterns and Process*. London: Blackwell, pp.240–268.

van der Pers JC, Loftstedt C (1983) Continuous single sensillum recording as a detection method for moth pheromone components in the effluent of a gas chromatograph. Physiol Ent 8:203–211.

van Lenteren JC, Bakker K, van Alphen JJM (1978) How to analyse host discrimination. Ecol Ent 3:71–75.

van Steenwyk RA, Ballmer GR, Page AL, Ganje TJ, Reynolds HT (1978) Dispersal of rubidium-marked pink bollworm. Env Ent 7:608–613.

Vargas R, Nishida T (1980) Life table of the corn earworm, *Heliothis zea* (Boddie), in sweet corn in Hawaii. Proc Hawaiian Ent Soc 23:301–307.

Varley GC, Gradwell GR (1960) Key factors in population studies. J Anim Ecol 29:399–401.

Varley GC, Gradwell GR, Hassell MP (1973) *Insect Population Ecology: An Analytical Approach*. Oxford: Blackwell Scientific.

Verheijen FJ (1960) The mechansims of the trapping effect of artificial light sources upon animals. Arch Neerland Zool 13:1–107.

Wagner TL, Wu H, Sharpe PJH, Coulson RN (1984a) Modelling distributions of insect development time:A literature review and application of the Weibull function. Ann Ent Soc Am 77:475–487.

Wagner TL, Wu H, Sharpe PJH, Schoolfield RM, Coulson RN (1984b) Modelling insect development rates: A literature review and application of a biophysical model. Ann Ent Soc Am 77:208–225.

Walker TJ (1985) Permanent traps for monitoring butterfly migration:tests in Florida, 1979–1984. J Lepid Soc 39:313–320.

Wang Y, Gutierrez AP, Oster G, Daxl R (1977) A population model for plant growth and development: Coupling cotton-herbivore interactions. Can Ent 109:1359–1374.

Waples RS (1989) Temporal variation in allele frequencies: Testing the right hypothesis. Evolution 43:1236–1251.

Wardhaugh KG, Room PM, Greenup LR (1980) The incidence of *Heliothis armigera* (Hübner) and *H. punctigera* Wallengren (Lepidoptera:Noctuidae) on cotton and other host plants in the Namoi valley of New South Wales. Bull Ent Res 70:113–131.

Weir BS, Cockerham CC (1984) Estimating F-statistics for the analysis of populations structure. Evolution 38:1358–1370.

Weiser J, Briggs JD (1971) Identification of pathogens. In: HD Burges, NW Hussey (Eds), *Microbial Control of Insects and Mites*. London: Academic Press,pp 13–66.

Wellington WG (1962) Population quality and the maintenance of nuclear polyhedrosis between outbreaks of *Malacosoma pluviale* (Dyar). J Insect Path 4:285–305.

Wellington WG, Cameron PJ, Thompson WA, Vertinsky IB, Landsberg AS (1975) A stochastic model for assessing the effects of external and internal heterogeneity on an insect population. Res Pop Ecol 17:1–28.

Wellso SG, Adkisson PL (1966) A long-day short-day effect in the photoperiodic control of the pupal diapause of the bollworm, *Heliothis zea* (Boddie). J Insect Physiol 12:1455–1465.

White EG (1964) A design for the effective killing of insects caught in light traps. NZ Entomol 3:25–27.

Whitlock VH (1974) Symptomatology of two viruses infecting *Heliothis armigera*. J Invertebr Pathol 23:70–75.

Whitten MJ (1979) The use of genetically selected strains for pest replacement or suppression. In: MA Hoy, JJ McKelvey Jr (Eds), *Genetics in Relation to Insect Management*. New York: The Rockefeller Foundation, pp.31–40.

Whitten MJ, McKenzie JA (1982) The genetic basis for pesticide resistance. In: KE Lee (Ed), *Proceedings 3rd Aust Conf Grassl Invertebr Ecol*, Adelaide: South Australian Govt Printer,pp.1–16.

Whittle CP, Bellas TE, Horak M, Pinese B (1987) The sex pheromone and taxonomic status of *Homona spargotis* Meyrick sp. rev., an Australian pest species of the *Coffearia* group (Lepidoptera:Tortricidae: Tortricinae). J Aust Ent Soc 26: 169–179.

Wiklund C (1981) Generalist vs specialist oviposition behaviour in *Papilio machaon* (Lepidoptera) and functional aspects of the hierarchy of oviposition preferences. Oikos 36:163–170.

Williams CB (1935) The times of activity of certain nocturnal insects, chiefly Lepidoptera, as indicated by a light trap. Trans Roy Ent Soc Lond 83:523–555.

Willmer PG (1982) Microclimate and the environmental physiology of insects. Adv Insect Physiol 16:1–57.

Wilson AGL (1974) Resistance of *Heliothis armigera* to insecticides in the Ord River Irrigation Area, north-western Australia. J Econ Entomol 67:256–258.

Wilson AGL (1983) Abundance and mortality of overwintering *Heliothis* spp. J Aust Ent Soc 22:191–199.

Wilson AGL (1984) Evaluation of pheromone trap design and dispensers for monitoring *Heliothis punctigera* and *H. armigera*. In: *Proc 4th Aust Appl Entomol Res Conf.* Govt Printer, Adelaide, 1984, pp.74–81.

Wilson AGL, Bauer L (1986) Light and pheromone traps:their place in monitoring *Heliothis* abundance. In: *Proc Aust Cott Conf Surfers Paradise, 1986, ACGRA* Wee Waa, pp. 239–245.

Wilson AGL, Greenup LR (1977) The relative injuriousness of insect pests of cotton in the Namoi Valley, New South Wales. Aust J Ecol 2:319–328.

Wilson AGL, Morton R (1989) Some factors affecting the reliability of pheromone traps for measurement of the relative abundance of *Helicoverpa punctigera* (Wallengren) and *H. armigera* (Hübner) (Lepidoptera:Noctuidae). Bull Ent Res 79:265–273.

Wilson AGL, Lewis T, Cunningham RB (1979) Overwintering and spring emergence of *Heliothis armigera* (Hübner) (Lepidoptera:Noctuidae) in the Namoi Valley, New South Wales. Bull Ent Res 69:97–109.

Wilson LT, Gutierrez AP, Leigh TF (1980) Within-plant distribution of the immatures of *Heliothis zea* (Boddie) on cotton. Hilgardia 48:12–23.

Wilson LT, Room PM (1982) The relative efficiency and reliability of three methods for sampling arthropods in Australian cotton fields. J Aust Ent Soc 21:175–181.

Wilson LT, Room PM (1983) Clumping patterns of fruit and arthropods in cotton, with implications for binomial sampling. Env Ent 12:50–54.

Wilson LT, Room PM, Bourne AS (1983) Dispersion of arthropods, flower buds and fruit in cotton fields: Effects of population density and season on the fit probability distributions. J Aust Ent Soc 22:129–134.

Wilson LT, Waite GK (1982) Feeding pattern of Australian *Heliothis* on cotton. Env Ent 11:297–300.

Wolf WW, Sparks AN, Pair SD, Westbook JK, Truesdale FM (1986) Radar observations and collections of insects in the Gulf of Mexico. In: E Danthanarayana (Ed), *Insect flight: Dispersal and Migration*. Berlin: Springer-Verlag, pp. 221–234.

Wolfenbarger DA, Lukefahr MJ, Graham HM (1973) LD_{50} values of methyl parathion and endrin to tobacco budworms and bollworms collected in the Americas and hypothesis on the spread of resistance in these Lepidopterans to these insecticides. J Econ Entomol 66:211–216.

Woodhead S (1982) p-Hydroxybenzaldehyde in the surface wax of sorghum: Its importance in seedling resistance to acridids. Ent Exp et Appl 31:296–302.

Wright S (1978) *Evolution and the Genetics of Populatons. Variability Within and Among Natural Populations*. Vol. 4. Chicago: University of Chicago Press.

Wright WE, Nitikin MI (1964) A survey of insects of cotton in New South Wales. Aust J Agric Sci 27:178–179.

Zacharuk RJ (1980) Ultrastructure and function of insect chemosensilla. Ann Rev Entomol 25:27–47.

Zacharuk RJ (1984) Antennae and sensilla. In: GA Kerkut, LI Gilbert (Eds), *Comprehensive Insect Physiology, Biochemistry and Pharmacology*. Vol 6, Oxford: Pergamon Press, pp. 1–69.

Zalucki MP, Kitching RL (1982a) Movement patterns in *Danaus plexippus*. Behaviour 80:174–198.

Zalucki MP, Kitching RL (1982b) Component analysis and modelling of the movement process: The simulation of simple tracks. Res Pop Ecol 24:239–249.

Zalucki MP, Kitching RL, Abel D, Pearson J (1980) A novel device for tracking butterflies in the field. Ann Ent Soc Am 73:262–265.

Zalucki MP, Twine PH (Eds). (1986) Proceedings of the *Heliothis* Ecology Workshop 1985. Brisbane: Qld. DPI Publication.

Zalucki MP, Daglish G, Firempong S, Twine PH (1986) The biology and ecology of *Heliothis armigera* (Hübner) and *H. punctigera* Wallengren (Lepidoptera:Noctuidae) in Australia:What do we know? Aust J Zool 34:779–814.

Zalucki MP, Brower LP, Malcolm SB (1990) Oviposition by *Danaus plexippus* in relation to cardenolide content of three *Asclepias* species in the south eastern U.S.A. Ecol Ent 15:231–240.

Zhulidov AV, Poltavskii AN, Emets VM (1982) Method for studying the migrations of nocturnal lepidopterous insects in geochemically heterogeneous regions. (In Russian) Ekologiya No. 6:51–54. (Soviet J Ecol 13:398–401, 1983.)

Index

Springer Series in Experimental Entomology
Editor: T.A. Miller

Insect Neurophysiological Techniques
By T.A. Miller

Neuroanatomical Techniques
Edited by N.J. Strausfeld and T.A. Miller

Sampling Methods in Soybean Entomology
Edited by M. Kogan and D. Herzog

Neurohormonal Techniques in Insects
Edited by T.A. Miller

Cuticle Techniques in Arthropods
Edited by T.A. Miller

Functional Neuroanatomy
Edited by N.J. Strausfeld

Techniques in Pheromone Research
Edited by H.E. Hummel and T.A. Miller

Measurement of Ion Transport and Metabolic Rate in Insects
Edited by T.J. Bradley and T.A. Miller

Neurochemical Techniques in Insect Research
Edited by H. Breer and T.A. Miller

Methods for the Study of Pest *Diabrotica*
Edited by J.L. Krysan and T.A. Miller

Insect-Plant Interactions
Edited by J.R. Miller and T.A. Miller

Immunological Techniques in Insect Biology
Edited by L.I. Gilbert and T.A. Miller

***Heliothis:* Research Methods and Prospects**
Edited by M.P. Zalucki

DATE DUE

JUL 0 2 1991		
NOV 1 8 1991		
JUN 0 6 1994		
MAY 2 2 1994		
MAR 2 4 2011		